PROPERTY TABLES
FOR THERMAL FLUIDS ENGINEERING

SI AND U.S. CUSTOMARY UNITS

Stephen R. Turns

David R. Kraige

CAMBRIDGE
UNIVERSITY PRESS

CAMBRIDGE UNIVERSITY PRESS

Cambridge, New York, Melbourne, Madrid, Cape Town, Singapore, São Paulo, Delhi

Cambridge University Press
32 Avenue of the Americas, New York, NY 10013-2473, USA

www.cambridge.org
Information on this title:www.cambridge.org/9780521709224

First published 2007

Printed in the United States of America

A catalogue record for this book is available from the British Library.

ISBN 978-0-521-70922-4 paperback

Contents

APPENDIX J: RADIATION PROPERTIES OF SELECTED MATERIALS AND SUBSTANCES

APPENDIX K: MACH NUMBER RELATIONSHIPS FOR COMPRESSIBLE FLOW

APPENDIX L: PSYCHROMETRY CHART

APPENDIX M: PROPERTIES OF THE ATMOSPHERE AT HIGH ALTITUDE

PART TWO   U.S. CUSTOMARY UNITS

APPENDIX AE: THERMODYNAMIC PROPERTIES OF REFRIGERANT-134a

APPENDIX BE: THERMODYNAMIC PROPERTIES OF IDEAL GASES AND CARBON

APPENDIX CE: THERMODYNAMIC AND THERMO-PHYSICAL PROPERTIES OF AIR

APPENDIX DE: THERMODYNAMIC PROPERTIES OF H_2O

APPENDIX EE: VARIOUS THERMODYNAMIC DATA

APPENDIX FE: THERMO-PHYSICAL PROPERTIES OF SELECTED GASES AT 1 atm

APPENDIX GE: THERMO-PHYSICAL PROPERTIES OF SELECTED LIQUIDS

APPENDIX HE: THERMO-PHYSICAL PROPERTIES OF HYDROCARBON FUELS

APPENDIX IE: THERMO-PHYSICAL PROPERTIES OF
SELECTED SOLIDS

APPENDIX LE: PSYCHROMETRY CHART

APPENDIX ME: PROPERTIES OF THE ATMOSPHERE AT HIGH ALTITUDES

Preface

The tables in the booklet complement the property tables in the appendices to *Thermodynamics: Concepts and Applications* and *Thermal-Fluid Science: An Integrated Approach* by Stephen R. Turns. In addition to duplicating the SI tables in these books in both SI and US Customary units, the present booklet contains property data for the refrigerant R-134a and properties of the atmosphere at high altitudes.

Part One

SI Units, Appendices A–M

Appendix A
Thermodynamic Properties of Refrigerant-134a

Table A.1 Saturation Properties of Refrigerant R-134a: Temperature Increments

Temp., °C	Press., P_{sat}, MPa	Specific Volume, m³/kg Sat. Liquid, v_f	Sat. Vapor, v_g	Internal Energy, kJ/kg Sat. Liquid, u_f	Sat. Vapor, u_g	Enthalpy, kJ/kg Sat. Liquid, h_f	Evap., h_{fg}	Sat. Vapor, h_g	Entropy, kJ/kg·K Sat. Liquid, s_f	Sat. Vapor, s_g
−40	0.051209	0.00070537	0.36108	148.11	355.51	148.14	225.86	374.00	0.79561	1.7643
−36	0.062908	0.00071120	0.29771	153.14	357.81	153.18	223.36	376.54	0.81700	1.7588
−32	0.076658	0.00071719	0.24727	158.19	360.11	158.25	220.81	379.06	0.83814	1.7538
−28	0.092703	0.00072336	0.20680	163.28	362.40	163.34	218.23	381.57	0.85906	1.7492
−24	0.11130	0.00072970	0.17407	168.39	364.70	168.47	215.60	384.07	0.87975	1.7451
−22	0.12165	0.00073294	0.16006	170.96	365.84	171.05	214.27	385.32	0.89002	1.7432
−20	0.13273	0.00073623	0.14739	173.54	366.99	173.64	212.91	386.55	0.90025	1.7413
−18	0.14460	0.00073958	0.13592	176.12	368.13	176.23	211.56	387.79	0.91042	1.7396
−16	0.15728	0.00074297	0.12551	178.72	369.28	178.83	210.19	389.02	0.92054	1.7379
−12	0.18524	0.00074994	0.10744	183.93	371.55	184.07	207.39	391.46	0.94066	1.7348
−8	0.21693	0.00075714	0.092422	189.17	373.82	189.34	204.53	393.87	0.96060	1.7320
−4	0.25268	0.00076460	0.079866	194.45	376.07	194.65	201.60	396.25	0.98037	1.7294
0	0.29280	0.00077233	0.069309	199.77	378.31	200.00	198.60	398.60	1.0000	1.7271
2	0.31462	0.00077631	0.064663	202.45	379.42	202.69	197.08	399.77	1.0098	1.7260
4	0.33766	0.00078037	0.060385	205.13	380.53	205.40	195.52	400.92	1.0195	1.7250
6	0.36198	0.00078451	0.056443	207.83	381.63	208.11	193.95	402.06	1.0292	1.7240
8	0.38761	0.00078873	0.052804	210.53	382.73	210.84	192.36	403.20	1.0388	1.7230
10	0.41461	0.00079305	0.049442	213.25	383.82	213.58	190.74	404.32	1.0485	1.7221
12	0.44301	0.00079745	0.046332	215.98	384.90	216.33	189.10	405.43	1.0581	1.7212
14	0.47288	0.00080196	0.043451	218.71	385.98	219.09	187.44	406.53	1.0677	1.7204
16	0.50425	0.00080657	0.040780	221.46	387.05	221.87	185.74	407.61	1.0772	1.7196
18	0.53718	0.00081128	0.038301	224.23	388.11	224.66	184.03	408.69	1.0867	1.7188
20	0.57171	0.00081610	0.035997	227.00	389.17	227.47	182.28	409.75	1.0962	1.7180
22	0.60789	0.00082105	0.033854	229.79	390.21	230.29	180.50	410.79	1.1057	1.7173
24	0.64578	0.00082612	0.031858	232.59	391.25	233.12	178.70	411.82	1.1152	1.7166
26	0.68543	0.00083131	0.029998	235.40	392.28	235.97	176.87	412.84	1.1246	1.7159
28	0.72688	0.00083665	0.028263	238.23	393.29	238.84	175.00	413.84	1.1341	1.7152

30	0.77020	0.00084213	0.026642	241.07	394.30	241.72	173.10	414.82	1.1435	1.7145
32	0.81543	0.00084777	0.025126	243.93	395.29	244.62	171.16	415.78	1.1529	1.7138
34	0.86263	0.00085357	0.023708	246.80	396.27	247.54	169.18	416.72	1.1623	1.7131
36	0.91185	0.00085954	0.022380	249.69	397.24	250.48	167.17	417.65	1.1717	1.7124
38	0.96315	0.00086569	0.021135	252.60	398.19	253.43	165.12	418.55	1.1811	1.7118
40	1.0166	0.00087204	0.019966	255.52	399.13	256.41	163.02	419.43	1.1905	1.7111
42	1.0722	0.00087859	0.018868	258.46	400.05	259.41	160.87	420.28	1.1999	1.7103
44	1.1301	0.00088537	0.017837	261.42	400.96	262.43	158.68	421.11	1.2092	1.7096
48	1.2529	0.00089965	0.015951	267.41	402.71	268.53	154.16	422.69	1.2280	1.7081
52	1.3854	0.00091502	0.014276	273.47	404.37	274.74	149.41	424.15	1.2469	1.7064
56	1.5282	0.00093166	0.012782	279.64	405.94	281.06	144.41	425.47	1.2658	1.7045
60	1.6818	0.00094979	0.011444	285.91	407.38	287.50	139.13	426.63	1.2848	1.7024
70	2.1168	0.0010038	0.0086527	302.16	410.33	304.28	124.37	428.65	1.3332	1.6956
80	2.6332	0.0010773	0.0064483	319.55	411.83	322.39	106.42	428.81	1.3836	1.6850
90	3.2442	0.0011936	0.0046134	339.06	410.45	342.93	82.490	425.42	1.4390	1.6662
100	3.9724	0.0015357	0.0026809	367.20	397.03	373.30	34.380	407.68	1.5188	1.6109

Table A.2 Saturation Properties of Refrigerant R-134a: Pressure Increments

Press., MPa	Temp., T_{sat}, °C	Specific Volume, m³/kg		Internal Energy, kJ/kg		Enthalpy, kJ/kg			Entropy, kJ/kg·K	
		Sat. Liquid, v_f	Sat. Vapor, v_g	Sat. Liquid, u_f	Sat. Vapor, u_g	Sat. Liquid, h_f	Evap., h_{fg}	Sat. Vapor, h_g	Sat. Liquid, s_f	Sat. Vapor, s_g
0.06	−36.935	0.00070982	0.31123	151.96	357.27	152.00	223.94	375.94	0.81202	1.7601
0.08	−31.115	0.00071854	0.23755	159.31	360.61	159.37	220.25	379.62	0.84278	1.7528
0.1	−26.361	0.00072593	0.19256	165.37	363.34	165.44	217.16	382.60	0.86756	1.7475
0.12	−22.310	0.00073244	0.16214	170.56	365.67	170.65	214.47	385.12	0.88844	1.7435
0.14	−18.760	0.00073830	0.14015	175.14	367.70	175.24	212.08	387.32	0.90656	1.7402
0.16	−15.588	0.00074368	0.12349	179.25	369.51	179.37	209.90	389.27	0.92262	1.7376
0.18	−12.712	0.00074868	0.11042	183.00	371.15	183.13	207.89	391.02	0.93709	1.7353
0.2	−10.076	0.00075337	0.099877	186.45	372.64	186.60	206.02	392.62	0.95027	1.7334
0.24	−5.3653	0.00076202	0.083906	192.65	375.30	192.83	202.61	395.44	0.97364	1.7303
0.28	−1.2277	0.00076993	0.072360	198.14	377.62	198.35	199.54	397.89	0.99399	1.7278
0.32	2.4768	0.00077727	0.063611	203.09	379.69	203.34	196.70	400.04	1.0121	1.7257
0.36	5.8412	0.00078418	0.056744	207.61	381.54	207.90	194.07	401.97	1.0284	1.7240
0.4	8.9306	0.00079073	0.051207	211.79	383.24	212.11	191.61	403.72	1.0433	1.7226
0.5	15.735	0.00080595	0.041123	221.10	386.91	221.50	185.97	407.47	1.0759	1.7197
0.6	21.572	0.00081998	0.034300	229.19	389.99	229.68	180.89	410.57	1.1037	1.7175
0.7	26.713	0.00083320	0.029365	236.41	392.64	236.99	176.21	413.20	1.1280	1.7156
0.8	31.327	0.00084585	0.025625	242.97	394.96	243.65	171.81	415.46	1.1497	1.7140
0.9	35.526	0.00085811	0.022687	249.01	397.01	249.78	167.65	417.43	1.1695	1.7126
1	39.388	0.00087007	0.020316	254.63	398.85	255.50	163.66	419.16	1.1876	1.7113
1.2	46.315	0.00089351	0.016718	264.88	401.98	265.95	156.09	422.04	1.2201	1.7087
1.4	52.422	0.00091671	0.014110	274.12	404.54	275.40	148.90	424.30	1.2489	1.7062
1.6	57.906	0.00094010	0.012126	282.61	406.64	284.11	141.93	426.04	1.2748	1.7036
1.8	62.895	0.00096400	0.010562	290.52	408.35	292.26	135.10	427.36	1.2987	1.7007
2	67.481	0.00098877	0.0092915	297.98	409.70	299.95	128.33	428.28	1.3209	1.6976
2.5	77.577	0.0010569	0.0069403	315.20	411.65	317.84	111.17	429.01	1.3711	1.6881
3	86.203	0.0011413	0.0052813	331.28	411.49	334.70	92.640	427.34	1.4171	1.6748

Table A.3   Superheated Vapor

Table A.3A $P = 0.06$ MPa ($T_{sat} = -36.935°C$)

T, °C	v, m³/kg	u, kJ/kg	h, kJ/kg	s, kJ/kg·K
Sat.	0.311230	357.27	375.94	1.7601
−20	0.336080	368.74	388.91	1.8131
−10	0.350490	375.70	396.73	1.8434
0	0.364760	382.80	404.69	1.8730
10	0.378930	390.07	412.81	1.9022
20	0.39303	397.50	421.08	1.9310
30	0.40705	405.10	429.52	1.9593
40	0.42102	412.86	438.12	1.9872
50	0.43495	420.79	446.88	2.0147
60	0.44884	428.88	455.81	2.0419
70	0.46269	437.14	464.90	2.0688
80	0.47652	445.56	474.15	2.0954
90	0.49032	454.14	483.56	2.1217

Table A.3B $P = 0.10$ MPa ($T_{sat} = -26.361°C$)

T, °C	v, m³/kg	u, kJ/kg	h, kJ/kg	s, kJ/kg·K
Sat.	0.192560	363.34	382.60	1.7475
−20	0.198410	367.81	387.65	1.7677
−10	0.207430	374.89	395.64	1.7986
0	0.216300	382.10	403.73	1.8288
10	0.225060	389.45	411.95	1.8584
20	0.23373	396.94	420.31	1.8874
30	0.24233	404.59	428.82	1.9160
40	0.25088	412.40	437.49	1.9441
50	0.25938	420.37	446.30	1.9718
60	0.26784	428.49	455.28	1.9991
70	0.27626	436.78	464.41	2.0261
80	0.28466	445.23	473.70	2.0528
90	0.29303	453.84	483.14	2.0792

Table A.3C $P = 0.14$ MPa ($T_{sat} = -18.760°C$)

T, °C	v, m³/kg	u, kJ/kg	h, kJ/kg	s, kJ/kg·K
Sat.	0.140150	367.70	387.32	1.7402
−10	0.146060	374.05	394.50	1.7680
0	0.152630	381.37	402.74	1.7987
10	0.159080	388.81	411.08	1.8287
20	0.165440	396.37	419.53	1.8580
30	0.17172	404.08	428.12	1.8868
40	0.17795	411.93	436.84	1.9151
50	0.18412	419.94	445.72	1.9430
60	0.19025	428.10	454.74	1.9705
70	0.19635	436.42	463.91	1.9977
80	0.20243	444.90	473.24	2.0244
90	0.20847	453.53	482.72	2.0509
100	0.21450	462.32	492.35	2.0771

Table A.3D $P = 0.18$ MPa ($T_{sat} = -12.712°C$)

T, °C	v, m³/kg	u, kJ/kg	h, kJ/kg	s, kJ/kg·K
Sat.	0.110420	371.15	391.02	1.7353
−10	0.111900	373.17	393.31	1.7441
0	0.117220	380.62	401.72	1.7754
10	0.122400	388.15	410.18	1.8059
20	0.127480	395.79	418.73	1.8355
30	0.13248	403.55	427.40	1.8646
40	0.13742	411.46	436.19	1.8931
50	0.14230	419.51	445.13	1.9212
60	0.14715	427.71	454.20	1.9489
70	0.15196	436.06	463.41	1.9761
80	0.15674	444.56	472.78	2.0030
90	0.16149	453.22	482.29	2.0296
100	0.16622	462.03	491.95	2.0558

Table A.3E $P = 0.20$ MPa ($T_{sat} = -10.076°C$)

T, °C	v, m³/kg	u, kJ/kg	h, kJ/kg	s, kJ/kg·K
Sat.	0.099877	372.64	392.62	1.7334
−10	0.099915	372.70	392.68	1.7337
0	0.104810	380.23	401.20	1.7654
10	0.109550	387.82	409.73	1.7961
20	0.114190	395.49	418.33	1.8259
30	0.11874	403.29	427.04	1.8551
40	0.12323	411.22	435.87	1.8838
50	0.12766	419.29	444.83	1.9120
60	0.13206	427.51	453.92	1.9397
70	0.13642	435.88	463.16	1.9670
80	0.14074	444.39	472.54	1.9939
90	0.14505	453.06	482.07	2.0206
100	0.14933	461.88	491.75	2.0468

Table A.3F $P = 0.24$ MPa ($T_{sat} = -5.3653°C$)

T, °C	v, m³/kg	u, kJ/kg	h, kJ/kg	s, kJ/kg·K
Sat.	0.083906	375.30	395.44	1.7303
0	0.086170	379.43	400.11	1.7475
10	0.090262	387.13	408.79	1.7787
20	0.094233	394.89	417.51	1.8090
30	0.098118	402.75	426.30	1.8385
40	0.10193	410.74	435.20	1.8674
50	0.10570	418.86	444.22	1.8957
60	0.10942	427.11	453.37	1.9236
70	0.11310	435.51	462.65	1.9511
80	0.11675	444.06	472.08	1.9781
90	0.12038	452.75	481.64	2.0048
100	0.12398	461.59	491.35	2.0312

Table A.3G $P = 0.28$ MPa ($T_{sat} = -1.2277°C$)

T, °C	v, m³/kg	u, kJ/kg	h, kJ/kg	s, kJ/kg·K
Sat.	0.072360	377.62	397.89	1.7278
0	0.072819	378.59	398.98	1.7318
10	0.076460	386.42	407.83	1.7636
20	0.079966	394.27	416.67	1.7943
30	0.083378	402.21	425.55	1.8241
40	0.086719	410.25	434.53	1.8532
50	0.090003	418.41	443.61	1.8818
60	0.093242	426.71	452.82	1.9098
70	0.096443	435.14	462.14	1.9374
80	0.099612	443.71	471.61	1.9646
90	0.10275	452.43	481.20	1.9914
100	0.10587	461.30	490.94	2.0178
110	0.10897	470.31	500.82	2.0440
120	0.11206	479.47	510.85	2.0698

Table A.3H $P = 0.32$ MPa ($T_{sat} = 2.4768°C$)

T, °C	v, m³/kg	u, kJ/kg	h, kJ/kg	s, kJ/kg·K
Sat.	0.063611	379.69	400.04	1.7257
10	0.066088	385.69	406.83	1.7501
20	0.069252	393.64	415.80	1.7812
30	0.072313	401.65	424.79	1.8113
40	0.075299	409.75	433.85	1.8407
50	0.078226	417.96	443.00	1.8695
60	0.081106	426.30	452.25	1.8977
70	0.083947	434.77	461.63	1.9254
80	0.086755	443.37	471.13	1.9527
90	0.089536	452.11	480.77	1.9796
100	0.092293	461.00	490.54	2.0062
110	0.095030	470.03	500.44	2.0324
120	0.097749	479.21	510.49	2.0583

Table A.3I $P = 0.40$ MPa ($T_{sat} = 8.9306°C$)

T, °C	v, m³/kg	u, kJ/kg	h, kJ/kg	s, kJ/kg·K
Sat.	0.051207	383.24	403.72	1.7226
10	0.051506	384.12	404.72	1.7261
20	0.054214	392.32	414.01	1.7584
30	0.056797	400.50	423.22	1.7893
40	0.059293	408.73	432.45	1.8192
50	0.061724	417.05	441.74	1.8484
60	0.064104	425.47	451.11	1.8770
70	0.066443	434.01	460.58	1.9050
80	0.068748	442.67	470.17	1.9325
90	0.071023	451.47	479.88	1.9597
100	0.073275	460.41	489.72	1.9864
110	0.075505	469.48	499.68	2.0127
120	0.077717	478.69	509.78	2.0387
130	0.079914	488.05	520.02	2.0645
140	0.082097	497.55	530.39	2.0899

Table A.3J $P = 0.50$ MPa ($T_{sat} = 15.735°C$)

T, °C	v, m³/kg	u, kJ/kg	h, kJ/kg	s, kJ/kg·K
Sat.	0.041123	386.91	407.47	1.7197
20	0.042116	390.55	411.61	1.7339
30	0.044338	398.99	421.16	1.7659
40	0.046456	407.40	430.63	1.7967
50	0.048499	415.86	440.11	1.8265
60	0.050486	424.40	449.64	1.8555
70	0.052427	433.04	459.25	1.8839
80	0.054331	441.78	468.95	1.9118
90	0.056205	450.65	478.76	1.9392
100	0.058054	459.65	488.68	1.9661
110	0.059880	468.78	498.72	1.9927
120	0.061688	478.04	508.88	2.0189
130	0.063479	487.44	519.18	2.0447
140	0.065257	496.98	529.60	2.0703

Table A.3K $P = 0.60$ MPa ($T_{sat} = 21.572°C$)

T, °C	v, m³/kg	u, kJ/kg	h, kJ/kg	s, kJ/kg·K
Sat.	0.034300	389.99	410.57	1.7175
30	0.035984	397.37	418.96	1.7455
40	0.037865	406.01	428.73	1.7772
50	0.039659	414.63	438.43	1.8077
60	0.041389	423.30	448.13	1.8373
70	0.043070	432.04	457.88	1.8661
80	0.044710	440.87	467.70	1.8943
90	0.046319	449.82	477.61	1.9220
100	0.047900	458.88	487.62	1.9492
110	0.049459	468.06	497.74	1.9759
120	0.050998	477.37	507.97	2.0023
130	0.052520	486.82	518.33	2.0283
140	0.054027	496.39	528.81	2.0540
150	0.055522	506.11	539.42	2.0794
160	0.057006	515.96	550.16	2.1045

Table A.3L $P = 0.70$ MPa ($T_{sat} = 26.713°C$)

T, °C	v, m³/kg	u, kJ/kg	h, kJ/kg	s, kJ/kg·K
Sat.	0.029365	392.64	413.20	1.7156
30	0.029966	395.62	416.60	1.7269
40	0.031696	404.53	426.72	1.7598
50	0.033322	413.35	436.67	1.7910
60	0.034875	422.16	446.57	1.8212
70	0.036374	431.01	456.47	1.8505
80	0.037829	439.94	466.42	1.8791
90	0.039250	448.97	476.44	1.9070
100	0.040642	458.09	486.54	1.9345
110	0.042010	467.34	496.74	1.9615
120	0.043358	476.70	507.05	1.9880
130	0.044689	486.19	517.47	2.0142
140	0.046004	495.80	528.01	2.0400
150	0.047307	505.55	538.67	2.0655
160	0.048598	515.44	549.45	2.0907

Table A.3M $P = 0.80$ MPa $(T_{sat} = 31.327°C)$

T, °C	v, m³/kg	u, kJ/kg	h, kJ/kg	s, kJ/kg·K
Sat.	0.025625	394.96	415.46	1.7140
40	0.027036	402.97	424.59	1.7436
50	0.028547	412.00	434.84	1.7758
60	0.029974	420.97	444.95	1.8067
70	0.031340	429.96	455.03	1.8365
80	0.032659	438.99	465.12	1.8654
90	0.033942	448.10	475.25	1.8937
100	0.035193	457.30	485.45	1.9214
110	0.036420	466.60	495.74	1.9486
120	0.037626	476.02	506.12	1.9754
130	0.038813	485.55	516.60	2.0017
140	0.039985	495.21	527.20	2.0277
150	0.041144	504.99	537.91	2.0533
160	0.042291	514.91	548.74	2.0786
170	0.043427	524.96	559.7	2.1036
180	0.044554	535.14	570.78	2.1283

Table A.3N $P = 0.90$ MPa $(T_{sat} = 35.526°C)$

T, °C	v, m³/kg	u, kJ/kg	h, kJ/kg	s, kJ/kg·K
Sat.	0.022687	397.01	417.43	1.7126
40	0.023375	401.28	422.32	1.7283
50	0.024810	410.59	432.92	1.7616
60	0.026146	419.75	443.28	1.7932
70	0.027414	428.87	453.54	1.8236
80	0.028630	438.01	463.78	1.8530
90	0.029807	447.21	474.03	1.8816
100	0.030951	456.48	484.34	1.9096
110	0.032069	465.85	494.71	1.9370
120	0.033164	475.32	505.17	1.9640
130	0.034241	484.91	515.72	1.9905
140	0.035302	494.61	526.38	2.0166
150	0.036349	504.43	537.14	2.0423
160	0.037384	514.38	548.02	2.0678
170	0.038408	524.46	559.02	2.0929
180	0.039423	534.67	570.15	2.1177

Table A.3O $P = 1.00$ MPa $(T_{sat} = 39.388°C)$

T, °C	v, m³/kg	u, kJ/kg	h, kJ/kg	s, kJ/kg·K
Sat.	0.020316	398.85	419.16	1.7113
40	0.020407	399.45	419.86	1.7135
50	0.021796	409.09	430.88	1.7482
60	0.023068	418.46	441.53	1.7806
70	0.024262	427.74	452.00	1.8116
80	0.025399	437.00	462.40	1.8414
90	0.026493	446.30	472.79	1.8705
100	0.027552	455.65	483.21	1.8988
110	0.028584	465.09	493.67	1.9264
120	0.029593	474.62	504.21	1.9536
130	0.030582	484.25	514.83	1.9803
140	0.031554	494.00	525.55	2.0065
150	0.032512	503.86	536.37	2.0324
160	0.033458	513.84	547.30	2.0579
170	0.034392	523.95	558.35	2.0831
180	0.035318	534.19	569.51	2.1080

Table A.3P $P = 1.20$ MPa ($T_{sat} = 46.315°C$)

T, °C	v, m³/kg	u, kJ/kg	h, kJ/kg	s, kJ/kg·K
Sat.	0.016718	401.98	422.04	1.7087
50	0.017201	405.77	426.41	1.7223
60	0.018404	415.70	437.79	1.7570
70	0.019502	425.36	448.76	1.7895
80	0.020530	434.90	459.53	1.8204
90	0.021506	444.41	470.22	1.8502
100	0.022443	453.94	480.87	1.8792
110	0.023348	463.53	491.54	1.9074
120	0.024228	473.18	502.25	1.9350
130	0.025087	482.92	513.03	1.9621
140	0.025928	492.76	523.87	1.9886
150	0.026753	502.70	534.81	2.0148
160	0.027566	512.76	545.84	2.0405
170	0.028367	522.93	556.97	2.0660
180	0.029159	533.23	568.22	2.0911

Table A.3Q $P = 1.40$ MPa ($T_{sat} = 52.422°C$)

T, °C	v, m³/kg	u, kJ/kg	h, kJ/kg	s, kJ/kg·K
Sat.	0.014110	404.54	424.30	1.7062
60	0.015005	412.61	433.62	1.7345
70	0.016060	422.77	445.25	1.7689
80	0.017023	432.65	456.48	1.8012
90	0.017923	442.42	467.52	1.8320
100	0.018778	452.16	478.45	1.8617
110	0.019597	461.90	489.34	1.8905
120	0.020388	471.70	500.24	1.9186
130	0.021156	481.55	511.17	1.9460
140	0.021904	491.49	522.16	1.9730
150	0.022636	501.52	533.21	1.9994
160	0.023355	511.65	544.35	2.0254
170	0.024061	521.89	555.58	2.0510
180	0.024758	532.25	566.91	2.0763

Table A.3R $P = 1.60$ MPa ($T_{sat} = 57.906°C$)

T, °C	v, m³/kg	u, kJ/kg	h, kJ/kg	s, kJ/kg·K
Sat.	0.012126	406.64	426.04	1.7036
60	0.012373	409.04	428.84	1.7120
70	0.013430	419.91	441.40	1.7491
80	0.014362	430.24	453.22	1.7831
90	0.015216	440.32	464.66	1.8150
100	0.016015	450.29	475.91	1.8456
110	0.016773	460.22	487.06	1.8751
120	0.017500	470.16	498.16	1.9037
130	0.018201	480.15	509.27	1.9316
140	0.018882	490.19	520.40	1.9589
150	0.019546	500.32	531.59	1.9856
160	0.020194	510.53	542.84	2.0119
170	0.020830	520.84	554.17	2.0378
180	0.021456	531.26	565.59	2.0632

Appendix B
Thermodynamic Properties of Ideal Gases and Carbon

Tables B.1–B.13 Present values for $\bar{c}_p(T)$, $\bar{h}^\circ(T) - \bar{h}^\circ_{f,ref}$, $\bar{h}^\circ_f(T)$, $\bar{s}^\circ(T)$, and $\Delta\bar{g}^\circ_f(T)$ at standard reference state ($T = 298.15$ K, P = 1 atm) for various species of the C–H–O–N system (with ideal-gas values for gaseous species).

Note that enthalpy of formation and Gibbs function of formation for compounds are calculated from the elements as follows:

$$\bar{h}^\circ_{f,i}(T) = \bar{h}^\circ_i(T) - \sum_{j\,\text{elements}} v'_j \bar{h}^\circ_j(T),$$

$$\bar{g}^\circ_{f,i}(T) = \bar{g}^\circ_i(T) - \sum_{j\,\text{elements}} v'_j \bar{g}^\circ_j(T)$$

$$= \bar{h}^\circ_{f,i}(T) - T\bar{s}^\circ_i(T) - \sum_{j\,\text{elements}} v'_j[-T\bar{s}^\circ_j(T)].$$

Sources: Tables B.1–B.12 are from Key, R. J., Rupley, F. M., and Miller, J. A., "The Chemkin Thermodynamic Data Base," Sandia Report, SAND87-8215B, March 1991. Table B.13 is from Myers, G. E., *Engineering Thermodynamics,* Prentice-Hall, Englewood Cliffs, NJ, 1989. Table B.14 is from Key, R. J., et al., ibid.

Table B.1   CO (Molecular Weight = 28.010, Enthalpy of Formation at 298 K = −110,541 kJ/kmol)

T (K)	$\bar{c}_p$ (kJ/kmol·K)	$\bar{h}°(T) − \bar{h}_f°(298)$ (kJ/kmol)	$\bar{h}_f°(T)$ (kJ/kmol)	$\bar{s}°(T)$ (kJ/kmol·K)	$\Delta\bar{g}_f°(T)$ (kJ/kmol)
200	28.687	−2835	−111,308	186.018	−128,532
298	29.072	0	−110,541	197.548	−137,163
300	29.078	54	−110,530	197.728	−137,328
400	29.433	2979	−110,121	206.141	−146,332
500	29.857	5943	−110,017	212.752	−155,403
600	30.407	8955	−110,156	218.242	−164,470
700	31.089	12,029	−110,477	222.979	−173,499
800	31.860	15,176	−110,924	227.180	−182,473
900	32.629	18,401	−111,450	230.978	−191,386
1000	33.255	21,697	−112,022	234.450	−200,238
1100	33.725	25,046	−112,619	237.642	−209,030
1200	34.148	28,440	−113,240	240.595	−217,768
1300	34.530	31,874	−113,881	243.344	−226,453
1400	34.872	35,345	−114,543	245.915	−235,087
1500	35.178	38,847	−115,225	248.332	−243,674
1600	35.451	42,379	−115,925	250.611	−252,214
1700	35.694	45,937	−116,644	252.768	−260,711
1800	35.910	49,517	−117,380	254.814	−269,164
1900	36.101	53,118	−118,132	256.761	−277,576
2000	36.271	56,737	−118,902	258.617	−285948
2100	36.421	60,371	−119,687	260.391	−294,281
2200	36.553	64,020	−120,488	262.088	−302,576
2300	36.670	67,682	−121,305	263.715	−310,835
2400	36.774	71,354	−122,137	265.278	−319,057
2500	36.867	75,036	−122,984	266.781	−327,245
2600	36.950	78,727	−123,847	268.229	−335,399
2700	37.025	82,426	−124,724	269.625	−343,519
2800	37.093	86,132	−125,616	270.973	−351,606
2900	37.155	89,844	−126,523	272.275	−359,661
3000	37.213	93,562	−127,446	273.536	−367,684
3100	37.268	97,287	−128,383	274.757	−375,677
3200	37.321	101,016	−129,335	275.941	−383,639
3300	37.372	104,751	−130,303	277.090	−391,571
3400	37.422	108,490	−131,285	278.207	−399,474
3500	37.471	112,235	−132,283	279.292	−407,347
3600	37.521	115,985	−133,295	280.349	−415,192
3700	37.570	119,739	−134,323	281.377	−423,008
3800	37.619	123,499	−135,366	282.380	−430,796
3900	37.667	127,263	−136,424	283.358	−438,557
4000	37.716	131,032	−137,497	284.312	−446,291
4100	37.764	134,806	−138,585	285.244	−453,997
4200	37.810	138,585	−139,687	286.154	−461,677
4300	37.855	142,368	−140,804	287.045	−469,330
4400	37.897	146,156	−141,935	287.915	−476,957
4500	37.936	149,948	−143,079	288.768	−484,558
4600	37.970	153,743	−144,236	289.602	−492,134
4700	37.998	157,541	−145,407	290.419	−499,684
4800	38.019	161,342	−146,589	291.219	−507,210
4900	38.031	165,145	−147,783	292.003	−514,710
5000	38.033	168,948	−148,987	292.771	−522,186

Table B.2 CO_2 (Molecular Weight = 44.011, Enthalpy of Formation at 298 K = −393,546 kJ/kmol)

T (K)	$\bar{c}_p$ (kJ/kmol·K)	$\bar{h}°(T) - \bar{h}_f°(298)$ (kJ/kmol)	$\bar{h}_f°(T)$ (kJ/kmol)	$\bar{s}°(T)$ (kJ/kmol·K)	$\Delta\bar{g}_f°(T)$ (kJ/kmol)
200	32.387	−3423	−393,483	199.876	−394,126
298	37.198	0	−393,546	213.736	−394,428
300	37.280	69	−393,547	213.966	−394,433
400	41.276	4003	−393,617	225.257	−394,718
500	44.569	8301	−393,712	234.833	−394,983
600	47.313	12,899	−393,844	243.209	−395,226
700	49.617	17,749	−394,013	250.680	−395,443
800	51.550	22,810	−394,213	257.436	−395,635
900	53.136	28,047	−394433	263.603	−395,799
1000	54.360	33,425	−394,659	269.268	−395,939
1100	55.333	38,911	−394,875	274.495	−396,056
1200	56.205	44,488	−395,083	279.348	−396,155
1300	56.984	50,149	−395,287	283.878	−396,236
1400	57.677	55,882	−395,88	288.127	−396,301
1500	58.292	61,681	−395,691	292.128	−396,352
1600	58.836	67,538	−395,897	295.908	−396,389
1700	59.316	73,446	−396,110	299.489	−396,414
1800	59.738	79,399	−396,332	302.892	−396,425
1900	60.108	85,392	−396,564	306.132	−396,424
2000	60.433	91,420	−396,808	309.223	−396,410
2100	60.717	97,477	−397,065	312.179	−396,384
2200	60.966	103,562	−397,338	315.009	−396,346
2300	61.185	109,670	−397,626	317.724	−396,294
2400	61.378	115,798	−397,931	320.333	−396,230
2500	61.548	121,944	−398,253	322.842	−396,152
2600	61.701	128,107	−398,594	325.259	−396,061
2700	61.839	134,284	−398,952	327.590	−395,957
2800	61.965	140,474	−399,329	329.841	−395,840
2900	62.083	146,677	−399,725	332.018	−395,708
3000	62.194	152,891	−400,140	334.124	−395,562
3100	62.301	159,116	−400,573	336.165	−395,403
3200	62.406	165,351	−401,025	338.145	−395,229
3300	62.510	171,597	−401,495	340.067	−395,041
3400	62.614	177,853	−401,983	341.935	−394,838
3500	62.718	184,120	−402,489	343.751	−394,620
3600	62.825	190,397	−403,013	345.519	−394,388
3700	62.932	196,685	−403,553	347.242	−394,141
3800	63.041	202,983	−404,110	348.922	−393,879
3900	63.151	209,293	−404,684	350.561	−393,602
4000	63.261	215,613	−405,273	352.161	−393,311
4100	63.369	221,945	−405,878	353.725	−393,004
4200	63.474	228,287	−406,499	355.253	−392,683
4300	63.575	234,640	−407,135	356.748	−392,346
4400	63.669	241,002	−407,785	358.210	−391,995
4500	63.753	247,373	−408,451	359.642	−391,629
4600	63.825	253,752	−409,132	361.044	−391,247
4700	63.881	260,138	−409,828	362.417	−390,851
4800	63.918	266,528	−410,539	363.763	−390,440
4900	63.932	272,920	−411,267	365.081	−390,014
5000	63.919	279,313	−412,010	366.372	−389,572

Table B.3 H₂ (Molecular Weight = 2.016, Enthalpy of Formation at 298 K = 0 kJ/kmol)

T (K)	$\bar{c}_p$ (kJ/kmol·K)	$\bar{h}°(T) - \bar{h}_f°(298)$ (kJ/kmol)	$\bar{h}_f°(T)$ (kJ/kmol)	$\bar{s}°(T)$ (kJ/kmol·K)	$\Delta\bar{g}_f°(T)$ (kJ/kmol)
200	28.522	−2818	0	119.137	0
298	28.871	0	0	130.595	0
300	28.877	53	0	130.773	0
400	29.120	2954	0	139.116	0
500	29.275	5874	0	145.632	0
600	29.375	8807	0	150.979	0
700	29.461	11,749	0	155.514	0
800	29.581	14,701	0	159.455	0
900	29.792	17,668	0	162.950	0
1000	30.160	20,664	0	166.106	0
1100	30.625	23,704	0	169.003	0
1200	31.077	26,789	0	171.687	0
1300	31.516	29,919	0	174.192	0
1400	31.943	33,092	0	176.543	0
1500	32.356	36,307	0	178.761	0
1600	32.758	39,562	0	180.862	0
1700	33.146	42,858	0	182.860	0
1800	33.522	46,191	0	184.765	0
1900	33.885	49,562	0	186.587	0
2000	34.236	52,968	0	188.334	0
2100	34.575	56,408	0	190.013	0
2200	34.901	59,882	0	191.629	0
2300	35.216	63,388	0	193.187	0
2400	35.519	66,925	0	194.692	0
2500	35.811	70,492	0	196.148	0
2600	36.091	74,087	0	197.558	0
2700	36.361	77,710	0	198.926	0
2800	36.621	81,359	0	200.253	0
2900	36.871	85,033	0	201.542	0
3000	37.112	88,733	0	202.796	0
3100	37.343	92,455	0	204.017	0
3200	37.566	96,201	0	205.206	0
3300	37.781	99,968	0	206.365	0
3400	37.989	103,757	0	207.496	0
3500	38.190	107,566	0	208.600	0
3600	38.385	111,395	0	209.679	0
3700	38.574	115,243	0	210.733	0
3800	38.759	119,109	0	211.764	0
3900	38.939	122,994	0	212.774	0
4000	39.116	126,897	0	213.762	0
4100	39.291	130,817	0	214.730	0
4200	39.464	134,755	0	215.679	0
4300	39.636	138,710	0	216.609	0
4400	39.808	142,682	0	217.522	0
4500	39.981	146,672	0	218.419	0
4600	40.156	150,679	0	219.300	0
4700	40.334	154,703	0	220.165	0
4800	40.516	158,746	0	221.016	0
4900	40.702	162,806	0	221.853	0
5000	40.895	166,886	0	222.678	0

Table B.4   H (Molecular Weight = 1.008, Enthalpy of Formation at 298 K = 217,979 kJ/kmol)

T (K)	$\bar{c}_p$ (kJ/kmol·K)	$\bar{h}°(T) - \bar{h}_f°(298)$ (kJ/kmol)	$\bar{h}_f°(T)$ (kJ/kmol)	$\bar{s}°(T)$ (kJ/kmol·K)	$\Delta\bar{g}_f°(T)$ (kJ/kmol)
200	20.786	−2040	217,346	106.305	207,999
298	20.786	0	217,977	114.605	203,276
300	20.786	38	217,989	114.733	203,185
400	20.786	2117	218,617	120.713	198,155
500	20.786	4196	219,236	125.351	192,968
600	20.786	6274	219,848	129.141	187,657
700	20.786	8353	220,456	132.345	182,244
800	20.786	10,431	221,059	135.121	176,744
900	20.786	12,510	221,653	137.569	171,169
1000	20.786	14,589	222,234	139.759	165,528
1100	20.786	16,667	222,793	141.740	159,830
1200	20.786	18,746	223,329	143.549	154,082
1300	20.786	20,824	223,843	145.213	148,291
1400	20.786	22,903	224,335	146.753	142,461
1500	20.786	24,982	224,806	148.187	136,596
1600	20.786	27,060	225,256	149.528	130,700
1700	20.786	29,139	225,687	150.789	124,777
1800	20.786	31,217	226,099	151.977	118,830
1900	20.786	33,296	226,493	153.101	112,859
2000	20.786	35,375	226,868	154.167	106,869
2100	20.786	37,453	227,226	155.181	100,860
2200	20.786	39,532	227,568	156.148	94,834
2300	20.786	41,610	227,894	157.072	88,794
2400	20.786	43,689	228,204	157.956	82,739
2500	20.786	45,768	228,499	158.805	76,672
2600	20.786	47,846	228,780	159.620	70,593
2700	20.786	49,925	229,047	160.405	64,504
2800	20.786	52,003	229,301	161.161	58,405
2900	20.786	54,082	229,543	161.890	52,298
3000	20.786	56,161	229,772	162.595	46,182
3100	20.786	58,239	229,989	163.276	40,058
3200	20.786	60,318	230,195	163.936	33,928
3300	20.786	62,396	230,390	164.576	27,792
3400	20.786	64,475	230,574	165.196	21,650
3500	20.786	66,554	230,748	165.799	15,502
3600	20.786	68,632	230,912	166.384	9350
3700	20.786	70,711	231,067	166.954	3194
3800	20.786	72,789	231,212	167.508	−2967
3900	20.786	74,868	231,348	168.048	−9132
4000	20.786	76,947	231,475	168.575	−15,299
4100	20.786	79,025	231,594	169.088	−21,470
4200	20.786	81,104	231,704	169.589	−27,644
4300	20.786	83,182	231,805	170.078	−33,820
4400	20.786	85,261	231,897	170.556	−39,998
4500	20.786	87,340	231,981	171.023	−46,179
4600	20.786	89,418	232,056	171.480	−52,361
4700	20.786	91,497	232,123	171.927	−58,545
4800	20.786	93,575	232,180	172.364	−64,730
4900	20.786	95,654	232,228	172.793	−70,916
5000	20.786	97,733	232,267	173.213	−77,103

Table B.5 OH (Molecular Weight = 17.007, Enthalpy of Formation at 298 K = 38,986 kJ/kmol)

T (K)	$\bar{c}_p$ (kJ/kmol·K)	$\bar{h}°(T) - \bar{h}_f°(298)$ (kJ/kmol)	$\bar{h}_f°(T)$ (kJ/kmol)	$\bar{s}°(T)$ (kJ/kmol·K)	$\Delta\bar{g}_f°(T)$ (kJ/kmol)
200	30.140	−2948	38,864	171.607	35,808
298	29.932	0	38,985	183.604	34,279
300	29.928	55	38,987	183.789	34,250
400	29.718	3037	39,030	192.369	32,662
500	29.570	6001	39,000	198.983	31,072
600	29.527	8955	38,909	204.369	29,494
700	29.615	11,911	38,770	208.925	27,935
800	29.844	14,883	38,599	212.893	26,399
900	30.208	17,884	38,410	216.428	24,885
1000	30.682	20,928	38,220	219.635	23,392
1100	31.186	24,022	38,039	222.583	21,918
1200	31.662	27,164	37,867	225.317	20,460
1300	32.114	30,353	37,704	227.869	19,017
1400	32.540	33,586	37,548	230.265	17,585
1500	32.943	36,860	37,397	232.524	16,164
1600	33.323	40,174	37,252	234.662	14,753
1700	33.682	43,524	37,109	236.693	13,352
1800	34.019	46,910	36,969	238.628	11,958
1900	34.337	50,328	36,831	240.476	10,573
2000	34.635	53,776	36,693	242.245	9194
2100	34.915	57,254	36,555	243.942	7823
2200	35.178	60,759	36,416	245.572	6458
2300	35.425	64,289	36,276	247.141	5099
2400	35.656	67,843	36,133	248.654	3746
2500	35.872	71,420	35,986	250.114	2400
2600	36.074	75,017	35,836	251.525	1060
2700	36.263	78,634	35,682	252.890	−275
2800	36.439	82,269	35,524	254.212	−1604
2900	36.604	85,922	35,360	255.493	−2927
3000	36.759	89,590	35,191	256.737	−4245
3100	36.903	93,273	35,016	257.945	−5556
3200	37.039	96,970	34,835	259.118	−6862
3300	37.166	100,681	34,648	260.260	−8162
3400	37.285	104,403	34,454	261.371	−9457
3500	37.398	108,137	34,253	262.454	−10,745
3600	37.504	111,882	34,046	263.509	−12,028
3700	37.605	115,638	33,831	264.538	−13,305
3800	37.701	119,403	33,610	265.542	−14,576
3900	37.793	123,178	33,381	266.522	−15,841
4000	37.882	126,962	33,146	267.480	−17,100
4100	37.968	130,754	32,903	268.417	−18,353
4200	38.052	134,555	32,654	269.333	−19,600
4300	38.135	138,365	32,397	270.229	−20,841
4400	38.217	142,182	32,134	271.107	−22,076
4500	38.300	146,008	31,864	271.967	−23,306
4600	38.382	149,842	31,588	272.809	−24,528
4700	38.466	153,685	31,305	273.636	−25,745
4800	38.552	157,536	31,017	274.446	−26,956
4900	38.640	161,395	30,722	275.242	−28,161
5000	38.732	165,264	30,422	276.024	−29,360

Table B.6 H$_2$O (Molecular Weight = 18.016, Enthalpy of Formation at 298 K = −241,845 kJ/kmol, Enthalpy of Vaporization = 44,010 kJ/kmol)

T (K)	$\bar{c}_p$ (kJ/kmol·K)	$\bar{h}°(T) - \bar{h}_f°(298)$ (kJ/kmol)	$\bar{h}_f°(T)$ (kJ/kmol)	$\bar{s}°(T)$ (kJ/kmol·K)	$\Delta\bar{g}_f°(T)$ (kJ/kmol)
200	32.255	−3227	−240,838	175.602	−232,779
298	33.448	0	−241,847	188.715	−228,608
300	33.468	62	−241,865	188.922	−228,526
400	34.437	3458	−242,858	198.686	−223,929
500	35.337	6947	−243,822	206.467	−219,085
600	36.288	10,528	−244,753	212.992	−214,049
700	37.364	14,209	−245,638	218.665	−208,861
800	38.587	18,005	−246,461	223.733	−203,550
900	39.930	21,930	−247,209	228.354	−198,141
1000	41.315	25,993	−247,879	232.633	−192,652
1100	42.638	30,191	−248,475	236.634	−187,100
1200	43.874	34,518	−249,005	240.397	−181,497
1300	45.027	38,963	−249,477	243.955	−175,852
1400	46.102	43,520	−249,895	247.332	−170,172
1500	47.103	48,181	−250,267	250.547	−164,464
1600	48.035	52,939	−250,597	253.617	−158,733
1700	48.901	57,786	−250,890	256.556	−152,983
1800	49.705	62,717	−251,151	259.374	−147,216
1900	50.451	67,725	−251,384	262.081	−141,435
2000	51.143	72,805	−251,594	264.687	−135,643
2100	51.784	77,952	−251,783	267.198	−129,841
2200	52.378	83,160	−251,955	269.621	−124,030
2300	52.927	88,426	−252,113	271.961	−118,211
2400	53.435	93,744	−252,261	274.225	−112,386
2500	53.905	99,112	−252,399	276.416	−106,555
2600	54.340	104,524	−252,532	278.539	−100,719
2700	54.742	109,979	−252,659	280.597	−94,878
2800	55.115	115,472	−252,785	282.595	−89,031
2900	55.459	121,001	−252,909	284.535	−83,181
3000	55.779	126,563	−253,034	286.420	−77,326
3100	56.076	132,156	−253,161	288.254	−71,467
3200	56.353	137,777	−253,290	290.039	−65,604
3300	56.610	143,426	−253,423	291.777	−59,737
3400	56.851	149,099	−253,561	293.471	−53,865
3500	57.076	154,795	−253,704	295.122	−47,990
3600	57.288	160,514	−253,852	296.733	−42,110
3700	57.488	166,252	−254,007	298.305	−36,226
3800	57.676	172,011	−254,169	299.841	−30,338
3900	57.856	177,787	−254,338	301.341	−24,446
4000	58.026	183,582	−254,515	302.808	−18,549
4100	58.190	189,392	−254,699	304.243	−12,648
4200	58.346	195,219	−254,892	305.647	−6742
4300	58.496	201,061	−255,093	307.022	−831
4400	58.641	206,918	−255,303	308.368	5085
4500	58.781	212,790	−255,522	309.688	11,005
4600	58.916	218,674	−255,751	310.981	16,930
4700	59.047	224,573	−255,990	312.250	22,861
4800	59.173	230,484	−256,239	313.494	28,796
4900	59.295	236,407	−256,501	314.716	34,737
5000	59.412	242,343	−256,774	315.915	40,684

Table B.7   N$_2$ (Molecular Weight = 28.013, Enthalpy of Formation at 298 K = 0 kJ/kmol)

T (K)	$\bar{c}_p$ (kJ/kmol·K)	$\bar{h}°(T) - \bar{h}_f°(298)$ (kJ/kmol)	$\bar{h}_f°(T)$ (kJ/kmol)	$\bar{s}°(T)$ (kJ/kmol·K)	$\Delta\bar{g}_f°(T)$ (kJ/kmol)
200	28.793	−2841	0	179.959	0
298	29.071	0	0	191.511	0
300	29.075	54	0	191.691	0
400	29.319	2973	0	200.088	0
500	29.636	5920	0	206.662	0
600	30.086	8905	0	212.103	0
700	30.684	11,942	0	216.784	0
800	31.394	15,046	0	220.927	0
900	32.131	18,222	0	224.667	0
1000	32.762	21,468	0	228.087	0
1100	33.258	24,770	0	231.233	0
1200	33.707	28,118	0	234.146	0
1300	34.113	31,510	0	236.861	0
1400	34.477	34,939	0	239.402	0
1500	34.805	38,404	0	241.792	0
1600	35.099	41,899	0	244.048	0
1700	35.361	45,423	0	246.184	0
1800	35.595	48,971	0	248.212	0
1900	35.803	52,541	0	250.142	0
2000	35.988	56,130	0	251.983	0
2100	36.152	59,738	0	253.743	0
2200	36.298	63,360	0	255.429	0
2300	36.428	66,997	0	257.045	0
2400	36.543	70,645	0	258.598	0
2500	36.645	74,305	0	260.092	0
2600	36.737	77,974	0	261.531	0
2700	36.820	81,652	0	262.919	0
2800	36.895	85,338	0	264.259	0
2900	36.964	89,031	0	265.555	0
3000	37.028	92,730	0	266.810	0
3100	37.088	96,436	0	268.025	0
3200	37.144	100,148	0	269.203	0
3300	37.198	103,865	0	270.347	0
3400	37.251	107,587	0	271.458	0
3500	37.302	111,315	0	272.539	0
3600	37.352	115,048	0	273.590	0
3700	37.402	118,786	0	274.614	0
3800	37.452	122,528	0	275.612	0
3900	37.501	126,276	0	276.586	0
4000	37.549	130,028	0	277.536	0
4100	37.597	133,786	0	278.464	0
4200	37.643	137,548	0	279.370	0
4300	37.688	141,314	0	280.257	0
4400	37.730	145,085	0	281.123	0
4500	37.768	148,860	0	281.972	0
4600	37.803	152,639	0	282.802	0
4700	37.832	156,420	0	283.616	0
4800	37.854	160,205	0	284.412	0
4900	37.868	163,991	0	285.193	0
5000	37.873	167,778	0	285.958	0

Table B.8 N (Molecular Weight = 14.007, Enthalpy of Formation at 298 K = 472,629 kJ/kmol)

T (K)	$\bar{c}_p$ (kJ/kmol·K)	$\bar{h}°(T) - \bar{h}_f°(298)$ (kJ/kmol)	$\bar{h}_f°(T)$ (kJ/kmol)	$\bar{s}°(T)$ (kJ/kmol·K)	$\Delta \bar{g}_f°(T)$ (kJ/kmol)
200	20.790	−2040	472,008	144.889	461,026
298	20.786	0	472,628	153.189	455,504
300	20.786	38	472,640	153.317	455,398
400	20.786	2117	473,258	159.297	449,557
500	20.786	4196	473,864	163.935	443,562
600	20.786	6274	474,450	167.725	437,446
700	20.786	8353	475,010	170.929	431,234
800	20.786	10,431	475,537	173.705	424,944
900	20.786	12,510	476,027	176.153	418,590
1000	20.786	14,589	476,483	178.343	412,183
1100	20.792	16,668	476,911	180.325	405,732
1200	20.795	18,747	477,316	182.134	399,243
1300	20.795	20,826	477,700	183.798	392,721
1400	20.793	22,906	478,064	185.339	386,171
1500	20.790	24,985	478,411	186.774	379,595
1600	20.786	27,064	478,742	188.115	372,996
1700	20.782	29,142	479,059	189.375	366,377
1800	20.779	31,220	479,363	190.563	359,740
1900	20.777	33,298	479,656	191.687	353,086
2000	20.776	35,376	479,939	192.752	346,417
2100	20.778	37,453	480,213	193.766	339,735
2200	20.783	39,531	480,479	194.733	333,039
2300	20.791	41,610	480,740	195.657	326,331
2400	20.802	43,690	480,995	196.542	319,612
2500	20.818	45,771	481,246	197.391	312,883
2600	20.838	47,853	481,494	198.208	306,143
2700	20.864	49,938	481,740	198.995	299,394
2800	20.895	52,026	481,985	199.754	292,636
2900	20.931	54,118	482,230	200.488	285,870
3000	20.974	56,213	482,476	201.199	279,094
3100	21.024	58,313	482,723	201.887	272,311
3200	21.080	60,418	482,972	202.555	265,519
3300	21.143	62,529	483,224	203.205	258,720
3400	21.214	64,647	483,481	203.837	251,913
3500	21.292	66,772	483,742	204.453	245,099
3600	21.378	68,905	484,009	205.054	238,276
3700	21.472	71,048	484,283	205.641	231,447
3800	21.575	73,200	484,564	206.215	224,610
3900	21.686	75,363	484,853	206.777	217,765
4000	21.805	77,537	485,151	207.328	210,913
4100	21.934	79,724	485,459	207.868	204,053
4200	22.071	81,924	485,779	208.398	197,186
4300	22.217	84,139	486,110	208.919	190,310
4400	22.372	86,368	486,453	209.431	183,427
4500	22.536	88,613	486,811	209.936	176,536
4600	22.709	90,875	487,184	210.433	169,637
4700	22.891	93,155	487,573	210.923	162,730
4800	23.082	95,454	487,979	211.407	155,814
4900	23.282	97,772	488,405	211.885	148,890
5000	23.491	100,111	488,850	212.358	141,956

Table B.9 NO (Molecular Weight = 30.006, Enthalpy of Formation at 298 K = 90,297 kJ/kmol)

T (K)	$\bar{c}_p$ (kJ/kmol·K)	$\bar{h}°(T) - \bar{h}_f°(298)$ (kJ/kmol)	$\bar{h}_f°(T)$ (kJ/kmol)	$\bar{s}°(T)$ (kJ/kmol·K)	$\Delta\bar{g}_f°(T)$ (kJ/kmol)
200	29.374	−2901	90,234	198.856	87,811
298	29.728	0	90,297	210.652	86,607
300	29.735	55	90,298	210.836	86,584
400	30.103	3046	90,341	219.439	85,340
500	30.570	6079	90,367	226.204	84,086
600	31.174	9165	90,382	231.829	82,828
700	31.908	12,318	90,393	236.688	81,568
800	32.715	15,549	90,405	241.001	80,307
900	33.489	18,860	90,421	244.900	79,043
1000	34.076	22,241	90,443	248.462	77,778
1100	34.483	25,669	90,465	251.729	76,510
1200	34.850	29,136	90,486	254.745	75,241
1300	35.180	32,638	90,505	257.548	73,970
1400	35.474	36,171	90,520	260.166	72,697
1500	35.737	39,732	90,532	262.623	71,423
1600	35.972	43,317	90,538	264.937	70,149
1700	36.180	46,925	90,539	267.124	68,875
1800	36.364	50,552	90,534	269.197	67,601
1900	36.527	54,197	90,523	271.168	66,327
2000	36.671	57,857	90,505	273.045	65,054
2100	36.797	61,531	90,479	274.838	63,782
2200	36.909	65,216	90,447	276.552	62,511
2300	37.008	68,912	90,406	278.195	61,243
2400	37.095	72,617	90,358	279.772	59,976
2500	37.173	76,331	90,303	281.288	58,711
2600	37.242	80,052	90,239	282.747	57,448
2700	37.305	83,779	90,168	284.154	56,188
2800	37.362	87,513	90,089	285.512	54,931
2900	37.415	91,251	90,003	286.824	53,677
3000	37.464	94,995	89,909	288.093	52,426
3100	37.511	98,744	89,809	289.322	51,178
3200	37.556	102,498	89,701	290.514	49,934
3300	37.600	106,255	89,586	291.670	48,693
3400	37.643	110,018	89,465	292.793	47,456
3500	37.686	113,784	89,337	293.885	46,222
3600	37.729	117,555	89,203	294.947	44,992
3700	37.771	121,330	89,063	295.981	43,766
3800	37.815	125,109	88,918	296.989	42,543
3900	37.858	128,893	88,767	297.972	41,325
4000	37.900	132,680	88,611	298.931	40,110
4100	37.943	136,473	88,449	299.867	38,900
4200	37.984	140,269	88,283	300.782	37,693
4300	38.023	144,069	88,112	301.677	36,491
4400	38.060	147,873	87,936	302.551	35,292
4500	38.093	151,681	87,755	303.407	34,098
4600	38.122	155,492	87,569	304.244	32,908
4700	38.146	159,305	87,379	305.064	31,721
4800	38.162	163,121	87,184	305.868	30,539
4900	38.171	166,938	86,984	306.655	29,361
5000	38.170	170,755	86,779	307.426	28,187

Table B.10 NO$_2$ (Molecular Weight = 46.006, Enthalpy of Formation at 298 K = 33,098 kJ/kmol)

T (K)	$\bar{c}_p$ (kJ/kmol·K)	$\bar{h}°(T) - \bar{h}_f°(298)$ (kJ/kmol)	$\bar{h}_f°(T)$ (kJ/kmol)	$\bar{s}°(T)$ (kJ/kmol·K)	$\Delta\bar{g}_f°(T)$ (kJ/kmol)
200	32.936	−3432	33,961	226.016	45,453
298	36.881	0	33,098	239.925	51,291
300	36.949	68	33,085	240.153	51,403
400	40.331	3937	32,521	251.259	57,602
500	43.227	8118	32,173	260.578	63,916
600	45.737	12,569	31,974	268.686	70,285
700	47.913	17,255	31,885	275.904	76,679
800	49.762	22,141	31,880	282.427	83,079
900	51.243	27,195	31,938	288.377	89,476
1000	52.271	32,375	32,035	293.834	95,864
1100	52.989	37,638	32,146	298.850	102,242
1200	53.625	42,970	32,267	303.489	108,609
1300	54.186	48,361	32,392	307.804	114,966
1400	54.679	53,805	32,519	311.838	121,313
1500	55.109	59,295	32,643	315.625	127,651
1600	55.483	64,825	32,762	319.194	133,981
1700	55.805	70,390	32,873	322.568	140,303
1800	56.082	75,984	32,973	325.765	146,620
1900	56.318	81,605	33,061	328.804	152,931
2000	56.517	87,247	33,134	331.698	159,238
2100	56.685	92,907	33,192	334.460	165,542
2200	56.826	98,583	33,233	337.100	171,843
2300	56.943	104,271	33,256	339.629	178,143
2400	57.040	109,971	33,262	342.054	184,442
2500	57.121	115,679	33,248	344.384	190,742
2600	57.188	121,394	33,216	346.626	197,042
2700	57.244	127,116	33,165	348.785	203,344
2800	57.291	132,843	33,095	350.868	209,648
2900	57.333	138,574	33,007	352.879	215,955
3000	57.371	144,309	32,900	354.824	222,265
3100	57.406	150,048	32,776	356.705	228,579
3200	57.440	155,791	32,634	358.529	234,898
3300	57.474	161,536	32,476	360.297	241,221
3400	57.509	167,285	32,302	362.013	247,549
3500	57.546	173,038	32,113	363.680	253,883
3600	57.584	178,795	31,908	365.302	260,222
3700	57.624	184,555	31,689	366.880	266,567
3800	57.665	190,319	31,456	368.418	272,918
3900	57.708	196,088	31,210	369.916	279,276
4000	57.750	201,861	30,951	371.378	285,639
4100	57.792	207,638	30,678	372.804	292,010
4200	57.831	213,419	30,393	374.197	298,387
4300	57.866	219,204	30,095	375.559	304,772
4400	57.895	224,992	29,783	376.889	311,163
4500	57.915	230,783	29,457	378.190	317,562
4600	57.925	236,575	29,117	379.464	323,968
4700	57.922	242,367	28,761	380.709	330,381
4800	57.902	248,159	28,389	381.929	336,803
4900	57.862	253,947	27,998	383.122	343,232
5000	57.798	259,730	27,586	384.290	349,670

Table B.11 O_2 (Molecular Weight = 31.999, Enthalpy of Formation at 298 K = 0 kJ/kmol)

T (K)	$\overline{c}_p$ (kJ/kmol·K)	$\overline{h}°(T) - \overline{h}_f°(298)$ (kJ/kmol)	$\overline{h}_f°(T)$ (kJ/kmol)	$\overline{s}°(T)$ (kJ/kmol·K)	$\Delta\overline{g}_f°(T)$ (kJ/kmol)
200	28.473	−2836	0	193.518	0
298	29.315	0	0	205.043	0
300	29.331	54	0	205.224	0
400	30.210	3031	0	213.782	0
500	31.114	6097	0	220.620	0
600	32.030	9254	0	226.374	0
700	32.927	12,503	0	231.379	0
800	33.757	15,838	0	235.831	0
900	34.454	19,250	0	239.849	0
1000	34.936	22,721	0	243.507	0
1100	35.270	26,232	0	246.852	0
1200	35.593	29,775	0	249.935	0
1300	35.903	33,350	0	252.796	0
1400	36.202	36,955	0	255.468	0
1500	36.490	40,590	0	257.976	0
1600	36.768	44,253	0	260.339	0
1700	37.036	47,943	0	262.577	0
1800	37.296	51,660	0	264.701	0
1900	37.546	55,402	0	266.724	0
2000	37.788	59,169	0	268.656	0
2100	38.023	62,959	0	270.506	0
2200	38.250	66,773	0	272.280	0
2300	38.470	70,609	0	273.985	0
2400	38.684	74,467	0	275.627	0
2500	38.891	78,346	0	277.210	0
2600	39.093	82,245	0	278.739	0
2700	39.289	86,164	0	280.218	0
2800	39.480	90,103	0	281.651	0
2900	39.665	94,060	0	283.039	0
3000	39.846	98,036	0	284.387	0
3100	40.023	102,029	0	285.697	0
3200	40.195	106,040	0	286.970	0
3300	40.362	110,068	0	288.209	0
3400	40.526	114,112	0	289.417	0
3500	40.686	118,173	0	290.594	0
3600	40.842	122,249	0	291.742	0
3700	40.994	126,341	0	292.863	0
3800	41.143	130,448	0	293.959	0
3900	41.287	134,570	0	295.029	0
4000	41.429	138,705	0	296.076	0
4100	41.566	142,855	0	297.101	0
4200	41.700	147,019	0	298.104	0
4300	41.830	151,195	0	299.087	0
4400	41.957	155,384	0	300.050	0
4500	42.079	159,586	0	300.994	0
4600	42.197	163,800	0	301.921	0
4700	42.312	168,026	0	302.829	0
4800	42.421	172,262	0	303.721	0
4900	42.527	176,510	0	304.597	0
5000	42.627	180,767	0	305.457	0

Table B.12 O (Molecular Weight = 16.000, Enthalpy of Formation at 298 K = 249,197 kJ/kmol)

T (K)	$\bar{c}_p$ (kJ/kmol·K)	$\bar{h}^\circ(T) - \bar{h}_f^\circ(298)$ (kJ/kmol)	$\bar{h}_f^\circ(T)$ (kJ/kmol)	$\bar{s}^\circ(T)$ (kJ/kmol·K)	$\Delta\bar{g}_f^\circ(T)$ (kJ/kmol)
200	22.477	−2176	248,439	152.085	237,374
298	21.899	0	249,197	160.945	231,778
300	21.890	41	249,211	161.080	231,670
400	21.500	2209	249,890	167.320	225,719
500	21.256	4345	250,494	172.089	219,605
600	21.113	6463	251,033	175.951	213,375
700	21.033	8570	251,516	179.199	207,060
800	20.986	10,671	251,949	182.004	200,679
900	20.952	12,768	252,340	184.474	194,246
1000	20.915	14,861	252,698	186.679	187,772
1100	20.898	16,952	253,033	188.672	181,263
1200	20.882	19,041	253,350	190.490	174,724
1300	20.867	21,128	253,650	192.160	168,159
1400	20.854	23,214	253,934	193.706	161,572
1500	20.843	25,299	254,201	195.145	154,966
1600	20.834	27,383	254,454	196.490	148,342
1700	20.827	29,466	254,692	197.753	141,702
1800	20.822	31,548	254,916	198.943	135,049
1900	20.820	33,630	255,127	200.069	128,384
2000	20.819	35,712	255,325	201.136	121,709
2100	20.821	37,794	255,512	202.152	115,023
2200	20.825	39,877	255,687	203.121	108,329
2300	20.831	41,959	255,852	204.047	101,627
2400	20.840	44,043	256,007	204.933	94,918
2500	20.851	46,127	256,152	205.784	88,203
2600	20.865	48,213	256,288	206.602	81,483
2700	20.881	50,300	256,416	207.390	74,757
2800	20.899	52,389	256,535	208.150	68,027
2900	20.920	54,480	256,648	208.884	61,292
3000	20.944	56,574	256,753	209.593	54,554
3100	20.970	58,669	256,852	210.280	47,812
3200	20.998	60,768	256,945	210.947	41,068
3300	21.028	62,869	257,032	211.593	34,320
3400	21.061	64,973	257,114	212.221	27,570
3500	21.095	67,081	257,192	212.832	20,818
3600	21.132	69,192	257,265	213.427	14,063
3700	21.171	71,308	257,334	214.007	7307
3800	21.212	73,427	257,400	214.572	548
3900	21.254	75,550	257,462	215.123	−6212
4000	21.299	77,678	257,522	215.662	−12,974
4100	21.345	79,810	257,579	216.189	−19,737
4200	21.392	81,947	257,635	216.703	−26,501
4300	21.441	84,088	257,688	217.207	−33,267
4400	21.490	86,235	257,740	217.701	−40,034
4500	21.541	88,386	257,790	218.184	−46,802
4600	21.593	90,543	257,840	218.658	−53,571
4700	21.646	92,705	257,889	219.123	−60,342
4800	21.699	94,872	257,938	219.580	−67,113
4900	21.752	97,045	257,987	220.028	−73,886
5000	21.805	99,223	258,036	220.468	−80,659

Table B.13 C(s) (Graphite, Molecular Weight = 12.011, Enthalpy of Formation at 298 K = 0 kJ/kmol)

	SI Values			
T (K)	$\bar{c}_p$ (kJ/kmol·K)	$\bar{h}°(T) - \bar{h}_f°(298)$ (kJ/kmol)	$\bar{h}°$ (kJ/kmol)	$\bar{s}°$ (kJ/kmol·K)
166.6667	3.155	−780.255	−779.080	2.435
177.7778	3.656	−742.406	−741.231	2.655
188.8889	4.148	−699.042	−697.868	2.891
200	4.630	−650.268	−649.093	3.142
211.1111	5.103	−596.184	−595.009	3.405
222.2222	5.568	−536.892	−535.717	3.679
233.3333	6.023	−472.491	−471.316	3.961
244.4444	6.470	−403.079	−401.904	4.252
255.5556	6.908	−328.753	−327.579	4.549
266.6667	7.337	−249.609	−248.434	4.852
277.7778	7.758	−165.739	−164.565	5.160
288.8889	8.171	−77.239	−76.064	5.473
298.15	8.509	0.000	1.175	5.736
300	8.575	15.803	16.977	5.789
311.1111	8.972	113.294	114.468	6.108
322.2222	9.360	215.146	216.320	6.429
333.3333	9.741	321.271	322.446	6.753
344.4444	10.114	431.583	432.757	7.079
355.5556	10.479	545.996	547.170	7.406
366.6667	10.837	664.426	665.600	7.734
377.7778	11.187	786.790	787.965	8.062
388.8889	11.530	913.006	914.181	8.392
400	11.866	1042.994	1044.169	8.721
411.1111	12.195	1176.674	1177.849	9.051
422.2222	12.517	1313.968	1315.142	9.380
433.3333	12.832	1454.798	1455.972	9.709
444.4444	13.140	1599.088	1600.262	10.038
455.5556	13.441	1746.762	1747.937	10.366
466.6667	13.736	1897.748	1898.923	10.694
477.7778	14.024	2051.972	2053.146	11.020
488.8889	14.305	2209.361	2210.536	11.346
500	14.581	2369.845	2371.020	11.671
511.1111	14.850	2533.355	2534.529	11.994
522.2222	15.113	2699.820	2700.994	12.316
533.3333	15.370	2869.173	2870.348	12.637
544.4444	15.621	3041.347	3042.522	12.957
555.5556	15.866	3216.276	3217.451	13.275
566.6667	16.105	3393.895	3395.070	13.591
577.7778	16.338	3574.140	3575.314	13.906
588.8889	16.566	3756.947	3758.121	14.220
600	16.788	3942.253	3943.428	14.531
611.1111	17.005	4129.999	4131.173	14.841
622.2222	17.216	4320.122	4321.296	15.150
633.3333	17.422	4512.563	4513.738	15.456
644.4444	17.623	4707.263	4708.438	15.761
655.5556	17.818	4904.164	4905.339	16.064
666.6667	18.009	5103.209	5104.384	16.365
677.7778	18.194	5304.341	5305.516	16.664
688.8889	18.374	5507.504	5508.679	16.962

	SI Values			
$T\ (K)$	$\bar{c}_p$ (kJ/kmol·K)	$\bar{h}°(T) - \bar{h}_f°(298)$ (kJ/kmol)	$\bar{h}°$ (kJ/kmol)	$\bar{s}°$ (kJ/kmol·K)
700	18.550	5712.644	5713.819	17.257
711.1111	18.720	5919.707	5920.881	17.550
722.2222	18.886	6128.637	6129.812	17.842
733.333	19.047	6339.385	6340.559	18.132
744.444	19.204	6551.896	6553.070	18.419
755.5556	19.356	6766.120	6767.294	18.705
766.6667	19.503	6982.006	6983.180	18.988
777.7778	19.646	7199.504	7200.678	19.270
788.8889	19.784	7418.565	7419.739	19.550
800	19.918	7639.140	7640.315	19.827
811.111	20.048	7861.181	7862.356	20.103
822.222	20.174	8084.642	8085.816	20.377
833.333	20.295	8309.474	8310.648	20.648
844.444	20.412	8535.631	8536.806	20.918
855.5556	20.526	8763.069	8764.244	21.185
866.6667	20.635	8991.742	8992.917	21.451
877.7778	20.740	9221.605	9222.780	21.715
888.8889	20.841	9452.615	9453.789	21.976
900	20.938	9684.727	9685.902	22.236
911.111	21.032	9917.899	9919.074	22.493
922.222	21.122	10152.088	10153.263	22.749
933.333	21.207	10387.253	10388.428	23.002
944.444	21.290	10623.352	10624.526	23.253
955.5556	21.368	10860.343	10861.517	23.503
966.6667	21.443	11098.185	11099.360	23.750
977.7778	21.514	11336.840	11338.014	23.996
988.8889	21.582	11576.266	11577.440	24.239
1000	21.646	11816.424	11817.598	24.481
1011.111	21.708	12057.282	12058.456	24.720
1022.222	21.769	12298.823	12299.998	24.958
1033.333	21.830	12541.043	12542.217	25.194
1044.444	21.890	12783.933	12785.107	25.427
1055.556	21.949	13027.485	13028.660	25.659
1077.778	22.066	13516.549	13517.724	26.118
1100	22.180	14008.178	14009.352	26.569
1122.222	22.292	14502.313	14503.487	27.014
1144.444	22.401	14998.898	15000.073	27.452
1166.667	22.507	15497.880	15499.055	27.884
1188.889	22.611	15999.204	16000.379	28.310
1211.111	22.713	16502.817	16503.992	28.729
1233.333	22.813	17008.668	17009.843	29.143
1255.556	22.910	17516.706	17517.880	29.552
1277.778	23.005	18026.881	18028.055	29.954
1300	23.098	18539.144	18540.319	30.352
1322.222	23.189	19053.448	19054.623	30.744
1344.444	23.278	19569.747	19570.921	31.131
1366.667	23.364	20087.994	20089.168	31.514
1388.889	23.449	20608.144	20609.319	31.891
1411.111	23.532	21130.154	21131.329	32.264
1433.333	23.612	21653.981	21655.156	32.632

(continued)

Table B.13   (continued)

	SI Values			
T (K)	$\bar{c}_p$ (kJ/kmol·K)	$\bar{h}°(T) - \bar{h}_f°(298)$ (kJ/kmol)	$\bar{h}°$ (kJ/kmol)	$\bar{s}°$ (kJ/kmol·K)
1455.556	23.691	22179.584	22180.758	32.996
1477.778	23.769	22706.920	22708.095	33.356
1500	23.844	23235.951	23237.125	33.711
1522.222	23.917	23766.636	23767.811	34.062
1544.444	23.989	24298.938	24300.113	34.410
1566.667	24.060	24832.819	24833.994	34.753
1588.889	24.128	25368.243	25369.418	35.092
1611.111	24.195	25905.174	25906.349	35.428
1633.333	24.261	26443.578	26444.752	35.760
1655.556	24.325	26983.420	26984.594	36.088
1677.778	24.387	27524.667	27525.842	36.413
1700	24.448	28067.288	28068.462	36.734
1722.222	24.508	28611.250	28612.424	37.052
1744.444	24.566	29156.523	29157.698	37.366
1766.667	24.623	29703.078	29704.252	37.678
1788.889	24.679	30250.884	30252.059	37.986
1811.111	24.734	30799.915	30801.090	38.291
1833.333	24.787	31350.143	31351.317	38.593
1855.556	24.839	31901.540	31902.714	38.892
1877.778	24.890	32454.081	32455.255	39.188
1900	24.939	33007.740	33008.915	39.481
1922.222	24.988	33562.494	33563.669	39.771
1944.444	25.036	34118.318	34119.492	40.059
1966.667	25.082	34675.189	34676.364	40.343
1988.889	25.128	35233.085	35234.260	40.626
2011.111	25.173	35791.984	35793.159	40.905
2033.333	25.216	36351.865	36353.040	41.182
2055.556	25.259	36912.708	36913.883	41.456
2077.778	25.301	37474.493	37475.668	41.728
2100	25.342	38037.202	38038.376	41.997
2122.222	25.383	38600.815	38601.989	42.264
2144.444	25.422	39165.315	39166.489	42.529
2166.667	25.461	39730.684	39731.859	42.791
2188.889	25.499	40296.907	40298.082	43.051
2211.111	25.536	40863.967	40865.142	43.309
2233.333	25.573	41431.850	41433.024	43.565
2255.556	25.609	42000.539	42001.714	43.818
2277.778	25.644	42570.022	42571.196	44.069
2300	25.679	43140.284	43141.458	44.318
2322.222	25.713	43711.312	43712.487	44.565
2344.444	25.747	44283.094	44284.269	44.811
2366.667	25.780	44855.618	44856.793	45.054
2388.889	25.813	45428.872	45430.047	45.295
2411.111	25.845	46002.846	46004.020	45.534
2433.333	25.876	46577.528	46578.702	45.771
2455.556	25.908	47152.909	47154.084	46.006
2477.778	25.939	47728.979	47730.154	46.240
2500	25.969	48305.730	48306.905	46.472
2522.222	25.999	48883.153	48884.327	46.702
2544.444	26.029	49461.239	49462.413	46.930
2566.667	26.058	50039.980	50041.155	47.156

	SI Values			
$T\ (K)$	$\bar{c}_p$ (kJ/kmol·K)	$\bar{h}°(T) - \bar{h}_f°(298)$ (kJ/kmol)	$\bar{h}°$ (kJ/kmol)	$\bar{s}°$ (kJ/kmol·K)
2588.889	26.087	50619.371	50620.546	47.381
2611.111	26.116	51199.403	51200.578	47.604
2633.333	26.144	51780.071	51781.246	47.826
2655.556	26.172	52361.368	52362.543	48.045
2677.778	26.200	52943.289	52944.464	48.264
2700	26.228	53525.828	53527.003	48.480
2722.222	26.256	54108.981	54110.156	48.695
2777.778	26.324	55569.519	55570.694	49.227
2833.333	26.391	57033.803	57034.978	49.748
2888.889	26.457	58501.785	58502.960	50.262
2944.444	26.522	59973.428	59974.603	50.766
3000	26.588	61448.707	61449.881	51.263
3055.556	26.653	62927.604	62928.779	51.751
3111.111	26.718	64410.114	64411.288	52.232
3166.667	26.783	65896.235	65897.410	52.705
3222.222	26.848	67385.977	67387.151	53.172
3277.778	26.914	68879.352	68880.526	53.631
3333.333	26.979	70376.378	70377.553	54.084
3388.889	27.046	71877.079	71878.254	54.531
3444.444	27.113	73381.480	73382.654	54.971
3500	27.180	74889.607	74890.782	55.405
3555.556	27.248	76401.490	76402.665	55.834
3611.111	27.316	77917.158	77918.332	56.257
3666.667	27.385	79436.637	79437.812	56.674
3722.222	27.454	80959.954	80961.129	57.087
3777.778	27.524	82487.133	82488.307	57.494
3833.333	27.594	84018.191	84019.366	57.896
3888.889	27.664	85553.145	85554.319	58.294
3944.444	27.735	87092.002	87093.177	58.687
4000	27.805	88634.765	88635.940	59.075
4055.556	27.875	90181.429	90182.603	59.459
4111.111	27.945	91731.978	91733.153	59.839
4166.667	28.014	93286.390	93287.565	60.214
4222.222	28.083	94844.630	94845.805	60.586
4277.778	28.150	96406.652	96407.826	60.953
4333.333	28.217	97972.397	97973.571	61.317
4388.889	28.282	99541.793	99542.968	61.677
4444.444	28.345	101114.755	101115.929	62.033
4500	28.406	102691.179	102692.354	62.386
4555.556	28.465	104270.948	104272.123	62.735
4611.111	28.522	105853.926	105855.101	63.080
4666.667	28.575	107439.959	107441.133	63.422
4722.222	28.625	109028.873	109030.047	63.760
4777.778	28.672	110620.474	110621.648	64.095
4833.333	28.714	112214.547	112215.722	64.427
4888.889	28.752	113810.856	113812.030	64.755
4944.444	28.785	115409.139	115410.313	65.081
5000	28.813	117009.111	117010.286	65.402

Table B.14 Curve-Fit Coefficients for Thermodynamic Properties (C–H–O–N System):

$$\bar{c}_p/R_u = a_1 + a_2 T + a_3 T^2 + a_4 T^3 + a_5 T^4,$$

$$\bar{h}^\circ/R_u T = a_1 + \frac{a_2}{2}T + \frac{a_3}{3}T^2 + \frac{a_4}{4}T^3 + \frac{a_5}{5}T^4 + \frac{a_6}{T},$$

$$\bar{s}^\circ/R_u = a_1 \ln T + a_2 T + \frac{a_3}{2}T^2 + \frac{a_4}{3}T^3 + \frac{a_5}{4}T^4 + a_7$$

Species	T (K)	a_1	a_2	a_3	a_4	a_5	a_6	a_7
CO	1000–5000	0.03025078E+02	0.14426885E-02	-0.05630827E-05	0.10185813E-09	-0.06910951E-13	-0.14268350E+05	0.06108217E+02
	300–1000	0.03262451E+02	0.15119409E-02	-0.03881755E-04	0.05581944E-07	-0.02474951E-10	-0.14310539E+05	0.04848897E+02
CO$_2$	1000–5000	0.04453623E+02	0.03140168E-01	-0.12784105E-05	0.02393996E-08	-0.16690333E-13	-0.04896696E+06	-0.09553959E+01
	300–1000	0.02275724E+02	0.09922072E-01	-0.10409113E-04	0.06866686E-07	-0.02117280E-10	-0.04837314E+06	0.10188488E+02
H$_2$	1000–5000	0.02991423E+02	0.07000644E-02	-0.05633828E-06	-0.09231578E-10	0.15827519E-14	-0.08350340E+04	-0.13551101E+01
	300–1000	0.03298124E+02	0.08249441E-02	-0.08143015E-05	0.09475434E-09	0.04134872E-11	-0.10125209E+04	-0.03294094E+02
H	1000–5000	0.02500000E+02	0.00000000E+00	0.00000000E+00	0.00000000E+00	0.00000000E+00	0.02547162E+06	-0.04601176E+01
	300–1000	0.02500000E+02	0.00000000E+00	0.00000000E+00	0.00000000E+00	0.00000000E+00	0.02547162E+06	-0.04601176E+01
OH	1000–5000	0.02882730E+02	0.10139743E-02	-0.02276877E-05	0.02174683E-09	-0.05126305E-14	0.03886888E+05	0.05595712E+02
	300–1000	0.03637266E+02	0.01850910E-02	-0.16761646E-05	0.02387202E-07	-0.08431442E-11	0.03606781E+05	0.13588605E+01
H$_2$O	1000–5000	0.02672145E+02	0.03056293E-01	-0.08730260E-05	0.12009964E-09	-0.06391618E-13	-0.02989921E+06	0.06862817E+02
	300–1000	0.03386842E+02	0.03474982E-01	-0.06354696E-04	0.06968581E-07	-0.02506588E-10	-0.03020811E+06	0.02590232E+02
N$_2$	1000–5000	0.02926640E+02	0.14879768E-02	-0.05684760E-05	0.10097038E-09	-0.06753351E-13	-0.09227977E+04	0.05980528E+02
	300–1000	0.03298677E+02	0.14082404E-02	-0.03963222E-04	0.05641515E-07	-0.02444854E-10	-0.10208999E+04	0.03950372E+02
N	1000–5000	0.02450268E+02	0.10661458E-03	-0.07465337E-06	0.01879652E-08	-0.10259839E-14	0.05611604E+06	0.04448758E+02
	300–1000	0.02503071E+02	-0.02180018E-03	0.05420529E-06	-0.05647560E-09	0.02099004E-12	0.05609890E+06	0.04167566E+02
NO	1000–5000	0.03245435E+02	0.12691383E-02	-0.05015890E-05	0.09169283E-09	-0.06275419E-13	0.09800840E+05	0.06417293E+02
	300–1000	0.03376541E+02	0.12530634E-02	-0.03302750E-04	0.05217810E-07	-0.02446262E-10	0.09817961E+05	0.05829590E+02
NO$_2$	1000–5000	0.04682859E+02	0.02462429E-01	-0.10422585E-05	0.01976902E-08	-0.13917168E-13	0.02261292E+05	0.09885985E+01
	300–1000	0.02670600E+02	0.07838500E-01	-0.08063864E-04	0.06161714E-07	-0.02320150E-10	0.02896290E+05	0.11612071E+02
O$_2$	1000–5000	0.03697578E+02	0.06135197E-02	-0.12588420E-06	0.01775281E-09	-0.11364354E-14	-0.12339301E+04	0.03189165E+02
	300–1000	0.03212936E+02	0.11274864E-02	-0.05756150E-05	0.13138773E-08	-0.08768554E-11	-0.10052490E+04	0.06034737E+02
O	1000–5000	0.02542059E+02	-0.02755061E-03	-0.03102803E-07	0.04551067E-10	-0.04368051E-14	0.02923080E+06	0.04920308E+02
	300–1000	0.02946428E+02	-0.16381665E-02	0.02421031E-04	-0.16028431E-08	0.03890696E-11	0.02914764E+06	0.02963995E+02

Appendix C
Thermodynamic and Thermo-Physical
Properties of Air

Table C.1 Approximate Composition, Apparent Molecular Weight, and Gas Constant for Dry Air

Constituent	Mole %
N_2	78.08
O_2	20.95
Ar	0.93
CO_2	0.036
Ne, He, CH_4, others	0.003

$$\mathcal{M}_{air} = 28.97 \text{ kg/kmol}$$
$$R_{air} = 287.0 \text{ J/kg} \cdot \text{K}$$

Table C.2 Thermodynamic Properties of Air at 1 atm*

T (K)	h (kJ/kg)	u (kJ/kg)	$s°$ (kJ/kg·K)	c_p (kJ/kg·K)	c_v (kJ/kg·K)	$\gamma \, (= c_p/c_v)$
200	325.42	268.14	3.4764	1.007	0.716	1.406
210	335.49	275.32	3.5255	1.007	0.716	1.405
220	345.55	282.49	3.5723	1.006	0.716	1.405
230	355.62	289.67	3.6171	1.006	0.716	1.404
240	365.68	296.85	3.6599	1.006	0.716	1.404
250	375.73	304.02	3.7009	1.006	0.717	1.404
260	385.79	311.20	3.7404	1.006	0.717	1.403
270	395.85	318.38	3.7783	1.006	0.717	1.403
280	405.91	325.56	3.8149	1.006	0.717	1.403
290	415.97	332.74	3.8502	1.006	0.718	1.402
300	426.04	339.93	3.8844	1.007	0.718	1.402
310	436.11	347.12	3.9174	1.007	0.719	1.401
320	446.18	354.31	3.9494	1.008	0.719	1.401
330	456.26	361.51	3.9804	1.008	0.720	1.400
340	466.34	368.72	4.0105	1.009	0.721	1.400
350	476.43	375.94	4.0397	1.009	0.721	1.399
360	486.53	383.16	4.0682	1.010	0.722	1.399
370	496.64	390.39	4.0959	1.011	0.723	1.398
380	506.75	397.63	4.1228	1.012	0.724	1.398
390	516.88	404.89	4.1491	1.013	0.725	1.397

*Property values generated from NIST Database 23: REFPROP Version 7.0 (August 2002). Reference state: h (78.903 K) = 0.0; s (78.903 K) = 0.0.

(continued)

Table C.2 (continued)

T (K)	h (kJ/kg)	u (kJ/kg)	$s°$ (kJ/kg·K)	c_p (kJ/kg·K)	c_v (kJ/kg·K)	γ (= c_p/c_v)
400	527.02	412.15	4.1748	1.014	0.727	1.396
410	537.17	419.42	4.1999	1.016	0.728	1.395
420	547.33	426.71	4.2244	1.017	0.729	1.395
430	557.51	434.01	4.2483	1.018	0.731	1.394
440	567.70	441.33	4.2717	1.020	0.732	1.393
450	577.90	448.66	4.2947	1.021	0.734	1.392
460	588.13	456.01	4.3171	1.023	0.735	1.391
470	598.36	463.38	4.3392	1.025	0.737	1.390
480	608.62	470.76	4.3608	1.026	0.739	1.389
490	618.89	478.16	4.3819	1.028	0.741	1.388
500	629.18	485.58	4.4027	1.030	0.743	1.387
510	639.50	493.01	4.4231	1.032	0.745	1.386
520	649.83	500.47	4.4432	1.034	0.747	1.385
530	660.18	507.95	4.4629	1.036	0.749	1.384
540	670.55	515.45	4.4823	1.038	0.751	1.383
550	680.94	522.97	4.5014	1.040	0.753	1.382
560	691.35	530.51	4.5201	1.042	0.755	1.381
570	701.79	538.07	4.5386	1.045	0.757	1.380
580	712.25	545.65	4.5568	1.047	0.759	1.378
590	722.73	553.26	4.5747	1.049	0.762	1.377
600	733.23	560.89	4.5924	1.051	0.764	1.376
610	743.76	568.55	4.6098	1.054	0.766	1.375
620	754.30	576.22	4.6269	1.056	0.769	1.374
630	764.88	583.92	4.6438	1.058	0.771	1.373
640	775.47	591.65	4.6605	1.061	0.773	1.371
650	786.09	599.40	4.6770	1.063	0.776	1.370
660	796.74	607.17	4.6932	1.066	0.778	1.369
670	807.41	614.96	4.7093	1.068	0.781	1.368
680	818.10	622.78	4.7251	1.070	0.783	1.367
690	828.81	630.63	4.7408	1.073	0.786	1.366
700	839.55	638.50	4.7562	1.075	0.788	1.365
710	850.32	646.39	4.7715	1.078	0.790	1.364
720	861.11	654.30	4.7866	1.080	0.793	1.362
730	871.92	662.24	4.8015	1.082	0.795	1.361
740	882.76	670.21	4.8162	1.085	0.798	1.360
750	893.62	678.20	4.8308	1.087	0.800	1.359
760	904.50	686.21	4.8452	1.090	0.802	1.358
770	915.41	694.24	4.8595	1.092	0.805	1.357
780	926.34	702.30	4.8736	1.094	0.807	1.356
790	937.29	710.39	4.8875	1.097	0.809	1.355
800	948.27	718.49	4.9014	1.099	0.812	1.354
820	970.30	734.77	4.9285	1.104	0.816	1.352
840	992.41	751.15	4.9552	1.108	0.821	1.350
860	1014.62	767.61	4.9813	1.113	0.825	1.348
880	1036.91	784.16	5.0069	1.117	0.830	1.346
900	1059.29	800.80	5.0321	1.121	0.834	1.344
920	1081.76	817.52	5.0568	1.125	0.838	1.343
940	1104.30	834.32	5.0810	1.129	0.842	1.341
960	1126.93	851.21	5.1048	1.133	0.846	1.339
980	1149.64	868.18	5.1283	1.137	0.850	1.338

T (K)	h (kJ/kg)	u (kJ/kg)	$s°$ (kJ/kg·K)	c_p (kJ/kg·K)	c_v (kJ/kg·K)	γ ($= c_p/c_v$)
1000	1172.43	885.22	5.1513	1.141	0.854	1.336
1020	1195.29	902.34	5.1739	1.145	0.858	1.335
1040	1218.23	919.53	5.1962	1.149	0.861	1.333
1060	1241.23	936.80	5.2181	1.152	0.865	1.332
1080	1264.31	954.13	5.2397	1.156	0.868	1.331
1100	1287.46	971.53	5.2609	1.159	0.872	1.329
1120	1310.67	989.00	5.2818	1.162	0.875	1.328
1140	1333.95	1006.54	5.3024	1.165	0.878	1.327
1160	1357.29	1024.13	5.3227	1.169	0.881	1.326
1180	1380.69	1041.79	5.3427	1.172	0.884	1.325
1200	1404.15	1059.51	5.3624	1.175	0.887	1.324
1220	1427.67	1077.29	5.3819	1.177	0.890	1.323
1240	1451.25	1095.12	5.4010	1.180	0.893	1.322
1260	1474.88	1113.01	5.4199	1.183	0.896	1.321
1280	1498.56	1130.95	5.4386	1.186	0.898	1.320
1300	1522.30	1148.95	5.4570	1.188	0.901	1.319
1320	1546.09	1166.99	5.4751	1.191	0.903	1.318
1340	1569.92	1185.08	5.4931	1.193	0.906	1.317
1360	1593.81	1203.23	5.5108	1.195	0.908	1.316
1380	1617.74	1221.42	5.5282	1.198	0.911	1.315
1400	1641.71	1239.65	5.5455	1.200	0.913	1.315
1420	1665.74	1257.93	5.5625	1.202	0.915	1.314
1440	1689.80	1276.25	5.5793	1.204	0.917	1.313
1460	1713.91	1294.61	5.5960	1.206	0.919	1.312
1480	1738.05	1313.02	5.6124	1.208	0.921	1.312
1500	1762.24	1331.46	5.6286	1.210	0.923	1.311
1520	1786.46	1349.94	5.6447	1.212	0.925	1.310
1540	1810.73	1368.46	5.6605	1.214	0.927	1.310
1560	1835.03	1387.02	5.6762	1.216	0.929	1.309
1580	1859.36	1405.61	5.6917	1.218	0.931	1.309
1600	1883.73	1424.24	5.7070	1.219	0.932	1.308
1620	1908.14	1442.90	5.7222	1.221	0.934	1.307
1640	1932.58	1461.60	5.7372	1.223	0.936	1.307
1660	1957.05	1480.33	5.7520	1.224	0.937	1.306
1680	1981.55	1499.09	5.7667	1.226	0.939	1.306
1700	2006.08	1517.88	5.7812	1.227	0.940	1.305
1750	2067.54	1564.99	5.8168	1.231	0.944	1.304
1800	2129.19	1612.27	5.8516	1.235	0.947	1.303
1850	2190.99	1659.72	5.8854	1.238	0.951	1.302
1900	2252.96	1707.33	5.9185	1.241	0.954	1.301
1950	2315.08	1755.09	5.9508	1.244	0.957	1.300
2000	2377.34	1803.00	5.9823	1.247	0.959	1.299
2050	2439.73	1851.04	6.0131	1.249	0.962	1.298
2100	2502.26	1899.21	6.0432	1.252	0.965	1.298
2150	2564.90	1947.50	6.0727	1.254	0.967	1.297
2200	2627.67	1995.91	6.1016	1.256	0.969	1.296
2250	2690.54	2044.43	6.1298	1.259	0.971	1.296
2300	2753.53	2093.05	6.1575	1.261	0.974	1.295
2350	2816.61	2141.78	6.1846	1.263	0.976	1.294

Table C.3   Thermo-Physical Properties of Air

Table C.3A   Thermo-Physical Properties of Air (100–1000 K at 1 atm)*

T (K)	ρ (kg/m^3)	c_p (kJ/kg·K)	μ (μPa·s)	ν (m^2/s)	k (W/m·K)	α (m^2/s)	Pr
100	3.6043	1.0356	7.1551	1.985E-06	0.010116	2.710E-06	0.73245
120	2.9772	1.0211	8.4995	2.855E-06	0.011996	3.946E-06	0.72349
140	2.5403	1.0142	9.7899	3.854E-06	0.013802	5.357E-06	0.71940
160	2.2169	1.0105	11.029	4.975E-06	0.015540	6.936E-06	0.71723
180	1.9674	1.0084	12.221	6.212E-06	0.017213	8.677E-06	0.71593
200	1.7688	1.0071	13.370	7.559E-06	0.018829	1.057E-05	0.71508
220	1.6068	1.0063	14.479	9.011E-06	0.020392	1.261E-05	0.71449
240	1.4721	1.0059	15.552	1.056E-05	0.021908	1.480E-05	0.71407
260	1.3584	1.0058	16.592	1.221E-05	0.023381	1.711E-05	0.71376
280	1.2610	1.0061	17.601	1.396E-05	0.024817	1.956E-05	0.71354
300	1.1767	1.0066	18.582	1.579E-05	0.026220	2.214E-05	0.71339
320	1.1030	1.0075	19.536	1.771E-05	0.027594	2.483E-05	0.71330
340	1.0380	1.0087	20.465	1.972E-05	0.028944	2.764E-05	0.71324
360	0.98022	1.0103	21.372	2.180E-05	0.030272	3.057E-05	0.71323
380	0.92856	1.0122	22.256	2.397E-05	0.031583	3.361E-05	0.71326
400	0.88208	1.0144	23.121	2.621E-05	0.032880	3.675E-05	0.71331
420	0.84004	1.0169	23.966	2.853E-05	0.034164	3.999E-05	0.71340
440	0.80183	1.0198	24.794	3.092E-05	0.035437	4.334E-05	0.71352
460	0.76695	1.0230	25.605	3.339E-05	0.036702	4.678E-05	0.71366
480	0.73497	1.0264	26.400	3.592E-05	0.037960	5.032E-05	0.71384
500	0.70556	1.0301	27.180	3.852E-05	0.039212	5.395E-05	0.71403
520	0.67842	1.0340	27.946	4.119E-05	0.040458	5.767E-05	0.71425
540	0.65329	1.0382	28.698	4.393E-05	0.041699	6.148E-05	0.71449
560	0.62995	1.0424	29.438	4.673E-05	0.042935	6.538E-05	0.71475
580	0.60823	1.0469	30.166	4.960E-05	0.044167	6.936E-05	0.71503
600	0.58795	1.0514	30.883	5.253E-05	0.045395	7.343E-05	0.71532
620	0.56898	1.0561	31.589	5.552E-05	0.046618	7.758E-05	0.71562
640	0.55120	1.0608	32.284	5.857E-05	0.047837	8.181E-05	0.71593
660	0.53450	1.0656	32.970	6.168E-05	0.049050	8.612E-05	0.71626
680	0.51878	1.0704	33.646	6.486E-05	0.050259	9.051E-05	0.71659
700	0.50396	1.0752	34.313	6.809E-05	0.051462	9.497E-05	0.71693
720	0.48996	1.0800	34.972	7.138E-05	0.052659	9.951E-05	0.71727
740	0.47672	1.0848	35.623	7.473E-05	0.053851	1.041E-04	0.71761
760	0.46417	1.0896	36.266	7.813E-05	0.055036	1.088E-04	0.71796
780	0.45227	1.0943	36.901	8.159E-05	0.056215	1.136E-04	0.71831
800	0.44097	1.0989	37.529	8.511E-05	0.057388	1.184E-04	0.71866
820	0.43022	1.1035	38.151	8.868E-05	0.058553	1.233E-04	0.71901
840	0.41997	1.1081	38.765	9.230E-05	0.059712	1.283E-04	0.71936
860	0.41021	1.1125	39.374	9.599E-05	0.060863	1.334E-04	0.71971
880	0.40089	1.1169	39.976	9.972E-05	0.062007	1.385E-04	0.72006
900	0.39198	1.1212	40.573	1.035E-04	0.063143	1.437E-04	0.72041
920	0.38346	1.1254	41.164	1.074E-04	0.064271	1.489E-04	0.72075
940	0.37530	1.1295	41.749	1.112E-04	0.065392	1.543E-04	0.72109
960	0.36749	1.1335	42.329	1.152E-04	0.066505	1.597E-04	0.72143
980	0.35999	1.1374	42.904	1.192E-04	0.067610	1.651E-04	0.72176
1000	0.35279	1.1412	43.474	1.232E-04	0.068708	1.707E-04	0.72210

*Values generated from NIST Database 23: REFPROP Version 7.0 (August 2002).

Table C.3B   Thermo-Physical Properties of Air (1000–2300 K at 1 atm)*

T (K)	ρ (kg/m^3)	c_p (kJ/kg·K)	μ (μPa·s)	v (m^2/s)	k (W/m·K)	α (m^2/s)	Pr
1000	0.35281	1.1412	43.474	1.23E-04	0.068708	6.871E-06	0.72210
1100	0.32074	1.1590	46.258	1.44E-04	0.074077	7.408E-06	0.72372
1200	0.29402	1.1745	48.941	1.66E-04	0.079254	7.925E-06	0.72530
1300	0.27141	1.1881	51.539	1.90E-04	0.084248	8.425E-06	0.72683
1400	0.25202	1.1999	54.063	2.15E-04	0.089070	8.907E-06	0.72835
1500	0.23523	1.2103	56.524	2.40E-04	0.093733	9.373E-06	0.72986
1600	0.22053	1.2194	58.930	2.67E-04	0.098251	9.825E-06	0.73138
1700	0.20756	1.2274	61.286	2.95E-04	0.10263	1.026E-05	0.73293
1800	0.19603	1.2345	63.600	3.24E-04	0.10690	1.069E-05	0.73451
1900	0.18571	1.2409	65.877	3.55E-04	0.11105	1.111E-05	0.73614
2000	0.17643	1.2466	68.119	3.86E-04	0.11509	1.151E-05	0.73781
2100	0.16803	1.2517	70.332	4.19E-04	0.11904	1.190E-05	0.73954
2200	0.16039	1.2564	72.518	4.52E-04	0.12290	1.229E-05	0.74134
2300	0.15342	1.2607	74.681	4.87E-04	0.12668	1.267E-05	0.74320

*Values generated from NIST Database 23: REFPROP Version 7.0 (August 2002).

Appendix D
Thermodynamic Properties of H₂O

Table D.1   Saturation Properties of Water and Steam: Temperature Increments

T (K)	P (kPa)	v (m³/kg) sat. liquid	v (m³/kg) sat. vapor	u (kJ/kg) sat. liquid	u (kJ/kg) sat. vapor	h (kJ/kg) sat. liquid	h (kJ/kg) sat. vapor	s (kJ/kg·K) sat. liquid	s (kJ/kg·K) sat. vapor
273.16	0.611650	0.0010002	205.99	0	2374.9	0.00061	2500.9	0	9.1555
274	0.650030	0.0010002	194.43	3.5435	2376.1	3.5442	2502.5	0.012952	9.1331
275	0.698460	0.0010001	181.60	7.7590	2377.5	7.7597	2504.3	0.028309	9.1066
276	0.750070	0.0010001	169.71	11.971	2378.8	11.972	2506.1	0.043600	9.0804
277	0.805020	0.0010001	158.70	16.181	2380.2	16.182	2508.0	0.058825	9.0544
278	0.863500	0.0010001	148.48	20.388	2381.6	20.389	2509.8	0.073985	9.0287
279	0.925700	0.0010001	139.00	24.593	2383.0	24.594	2511.6	0.089083	9.0031
280	0.991830	0.0010001	130.19	28.795	2384.3	28.796	2513.4	0.10412	8.9779
281	1.062200	0.0010002	122.01	32.996	2385.7	32.997	2515.3	0.11909	8.9528
282	1.136800	0.0010003	114.40	37.194	2387.1	37.195	2517.1	0.13401	8.9280
283	1.216000	0.0010003	107.32	41.391	2388.4	41.392	2518.9	0.14886	8.9034
284	1.300000	0.0010004	100.74	45.586	2389.8	45.587	2520.8	0.16366	8.8791
285	1.389100	0.0010005	94.602	49.779	2391.2	49.780	2522.6	0.17840	8.8549
286	1.483600	0.0010006	88.887	53.971	2392.6	53.973	2524.4	0.19308	8.8310
287	1.583600	0.0010008	83.560	58.162	2393.9	58.163	2526.2	0.20771	8.8073
288	1.689500	0.0010009	78.592	62.351	2395.3	62.353	2528.1	0.22228	8.7838
289	1.801600	0.0010011	73.955	66.540	2396.7	66.542	2529.9	0.23680	8.7605
290	1.920100	0.0010012	69.625	70.727	2398.0	70.729	2531.7	0.25126	8.7374
291	2.045400	0.0010014	65.581	74.914	2399.4	74.916	2533.5	0.26568	8.7145
292	2.177900	0.0010016	61.801	79.099	2400.8	79.101	2535.3	0.28003	8.6918
293	2.317800	0.0010018	58.267	83.284	2402.1	83.286	2537.2	0.29434	8.6693
294	2.465500	0.0010020	54.960	87.468	2403.5	87.471	2539.0	0.30860	8.6471
295	2.621300	0.0010022	51.865	91.652	2404.8	91.654	2540.8	0.32280	8.6250
296	2.785700	0.0010025	48.966	95.835	2406.2	95.837	2542.6	0.33696	8.6031
297	2.959100	0.0010027	46.251	100.02	2407.6	100.02	2544.4	0.35106	8.5814
298	3.141800	0.0010030	43.705	104.20	2408.9	104.20	2546.2	0.36512	8.5599
299	3.334300	0.0010032	41.318	108.38	2410.3	108.38	2548.0	0.37913	8.5385
300	3.536900	0.0010035	39.078	112.56	2411.6	112.56	2549.9	0.39309	8.5174
301	3.750200	0.0010038	36.976	116.74	2413.0	116.75	2551.7	0.40700	8.4964
302	3.974600	0.0010041	35.002	120.92	2414.4	120.93	2553.5	0.42087	8.4756
303	4.210600	0.0010044	33.147	125.10	2415.7	125.11	2555.3	0.43469	8.4550
304	4.458700	0.0010047	31.403	129.28	2417.1	129.29	2557.1	0.44846	8.4346
305	4.719400	0.0010050	29.764	133.46	2418.4	133.47	2558.9	0.46219	8.4144
306	4.993299	0.0010053	28.222	137.64	2419.8	137.65	2560.7	0.47587	8.3943
307	5.280799	0.0010057	26.770	141.82	2421.1	141.83	2562.5	0.48950	8.3744
308	5.582599	0.0010060	25.403	146.00	2422.5	146.01	2564.3	0.50310	8.3546
309	5.899199	0.0010063	24.116	150.18	2423.8	150.19	2566.1	0.51664	8.3351
310	6.231199	0.0010067	22.903	154.36	2425.2	154.37	2567.9	0.53015	8.3156
311	6.579299	0.0010071	21.759	158.54	2426.5	158.55	2569.7	0.54361	8.2964
312	6.944099	0.0010074	20.680	162.72	2427.8	162.73	2571.5	0.55702	8.2773
313	7.326199	0.0010078	19.663	166.90	2429.2	166.91	2573.2	0.57040	8.2584
314	7.726299	0.0010082	18.702	171.08	2430.5	171.09	2575.0	0.58373	8.2396
315	8.145199	0.0010086	17.795	175.26	2431.9	175.27	2576.8	0.59702	8.2210
316	8.583499	0.0010090	16.938	179.44	2433.2	179.45	2578.6	0.61027	8.2025
317	9.041899	0.0010094	16.129	183.62	2434.5	183.63	2580.4	0.62348	8.1842
318	9.521299	0.0010099	15.363	187.80	2435.9	187.81	2582.2	0.63664	8.1660
319	10.022989	0.0010103	14.639	191.98	2437.2	191.99	2583.9	0.64977	8.1480
320	10.546989	0.0010107	13.954	196.16	2438.5	196.17	2585.7	0.66285	8.1302

T (K)	P (MPa)	v (m³/kg) sat. liquid	v (m³/kg) sat. vapor	u (kJ/kg) sat. liquid	u (kJ/kg) sat. vapor	h (kJ/kg) sat. liquid	h (kJ/kg) sat. vapor	s (kJ/kg·K) sat. liquid	s (kJ/kg·K) sat. vapor
320	0.010547	0.0010107	13.954	196.16	2438.5	196.17	2585.7	0.66285	8.1302
325	0.013532	0.0010130	11.039	217.07	2445.2	217.08	2594.6	0.72768	8.0430
330	0.017214	0.0010155	8.8050	237.98	2451.8	238.00	2603.3	0.79154	7.9592
335	0.021719	0.0010181	7.0788	258.9	2458.3	258.93	2612.1	0.85447	7.8787
340	0.027189	0.0010209	5.7339	279.84	2464.8	279.87	2620.7	0.91650	7.8013
345	0.033784	0.0010239	4.6776	300.79	2471.2	300.82	2629.3	0.97766	7.7267
350	0.041683	0.0010270	3.8419	321.75	2477.6	321.79	2637.7	1.0380	7.6549
355	0.051081	0.0010303	3.1759	342.73	2483.9	342.78	2646.1	1.0975	7.5857
360	0.062195	0.0010337	2.6414	363.73	2490.1	363.79	2654.4	1.1562	7.5190
365	0.075261	0.0010373	2.2098	384.75	2496.2	384.82	2662.5	1.2142	7.4545
370	0.090536	0.0010410	1.8590	405.79	2502.3	405.88	2670.6	1.2715	7.3923
375	0.108310	0.0010449	1.5722	426.86	2508.2	426.97	2678.5	1.3281	7.3321
380	0.128860	0.0010490	1.3364	447.96	2514.1	448.09	2686.2	1.3839	7.2738
385	0.152529	0.0010532	1.1414	469.09	2519.8	469.25	2693.9	1.4392	7.2174
390	0.179649	0.0010575	0.97928	490.25	2525.4	490.44	2701.3	1.4938	7.1627
395	0.210609	0.0010620	0.84389	511.45	2530.9	511.67	2708.6	1.5478	7.1097
400	0.245779	0.0010667	0.73024	532.69	2536.2	532.95	2715.7	1.6013	7.0581
405	0.285589	0.0010715	0.63441	553.98	2541.4	554.28	2722.6	1.6541	7.0081
410	0.330459	0.0010765	0.55323	575.31	2546.5	575.66	2729.3	1.7065	6.9593
415	0.380879	0.0010817	0.48418	596.69	2551.4	597.10	2735.8	1.7583	6.9119
420	0.437309	0.0010870	0.42520	618.13	2556.2	618.60	2742.1	1.8097	6.8656
425	0.500259	0.0010926	0.37463	639.62	2560.7	640.17	2748.1	1.8606	6.8205
430	0.570269	0.0010983	0.33110	661.18	2565.1	661.80	2753.9	1.9110	6.7764
435	0.647879	0.0011042	0.29350	682.8	2569.3	683.52	2759.5	1.9610	6.7333
440	0.733679	0.0011103	0.26090	704.5	2573.3	705.31	2764.7	2.0106	6.6911
445	0.828249	0.0011166	0.23255	726.26	2577.1	727.19	2769.7	2.0598	6.6498
450	0.932209	0.0011232	0.20781	748.11	2580.7	749.16	2774.4	2.1087	6.6092
455	1.046289	0.0011299	0.18616	770.05	2584.0	771.23	2778.8	2.1571	6.5694
460	1.170988	0.0011369	0.16715	792.07	2587.2	793.41	2782.9	2.2053	6.5303
465	1.306987	0.0011442	0.15041	814.2	2590.1	815.69	2786.6	2.2532	6.4917
470	1.455187	0.0011517	0.13564	836.42	2592.7	838.09	2790.0	2.3007	6.4538
475	1.616086	0.0011594	0.12255	858.75	2595.0	860.62	2793.1	2.3480	6.4164
480	1.790586	0.0011675	0.11094	881.19	2597.1	883.28	2795.8	2.3950	6.3794
485	1.979285	0.0011758	0.10061	903.76	2598.9	906.09	2798.1	2.4418	6.3428
490	2.183184	0.0011845	0.091390	926.45	2600.5	929.04	2800.0	2.4884	6.3066
495	2.402884	0.0011935	0.083149	949.28	2601.7	952.15	2801.4	2.5348	6.2708
500	2.639283	0.0012029	0.075764	972.26	2602.5	975.43	2802.5	2.5810	6.2351
505	2.893182	0.0012127	0.069131	995.38	2603.0	998.89	2803.1	2.6271	6.1997
510	3.165582	0.0012228	0.063161	1018.7	2603.2	1022.5	2803.2	2.6731	6.1645
515	3.457181	0.0012334	0.057776	1042.1	2603.0	1046.4	2802.7	2.7189	6.1293
520	3.769080	0.0012445	0.052910	1065.8	2602.4	1070.5	2801.8	2.7647	6.0942
525	4.101980	0.0012561	0.048503	1089.6	2601.4	1094.8	2800.3	2.8104	6.0591
530	4.456979	0.0012682	0.044503	1113.7	2599.9	1119.3	2798.2	2.8561	6.0239
535	4.834978	0.0012809	0.040868	1138	2597.9	1144.2	2795.5	2.9019	5.9885
540	5.236977	0.0012942	0.037556	1162.5	2595.5	1169.3	2792.2	2.9476	5.9530
545	5.664076	0.0013083	0.034535	1187.3	2592.5	1194.7	2788.1	2.9935	5.9171
550	6.117276	0.0013231	0.031772	1212.4	2588.9	1220.5	2783.3	3.0394	5.8809
555	6.597675	0.0013387	0.029242	1237.8	2584.8	1246.6	2777.7	3.0855	5.8443
560	7.106274	0.0013553	0.026920	1263.5	2579.9	1273.1	2771.2	3.1319	5.8071
565	7.644473	0.0013729	0.024786	1289.6	2574.4	1300.1	2763.9	3.1785	5.7693
570	8.213272	0.0013917	0.022820	1316	2568.0	1327.5	2755.5	3.2254	5.7307
575	8.814071	0.0014118	0.021005	1343	2560.9	1355.4	2746.0	3.2727	5.6912
580	9.448070	0.0014334	0.019328	1370.4	2552.7	1383.9	2735.3	3.3205	5.6506
585	10.117686	0.0014567	0.017773	1398.4	2543.5	1413.1	2723.3	3.3690	5.6087
590	10.821674	0.0014820	0.016329	1426.9	2533.2	1443.0	2709.9	3.4181	5.5654
595	11.563662	0.0015095	0.014984	1456.2	2521.5	1473.7	2694.8	3.4680	5.5203
600	12.345649	0.0015399	0.013728	1486.4	2508.3	1505.4	2677.8	3.5190	5.4731
605	13.167635	0.0015735	0.012552	1517.4	2493.4	1538.1	2658.7	3.5713	5.4234
610	14.033620	0.0016112	0.011446	1549.6	2476.4	1572.2	2637.0	3.6252	5.3707
615	14.943604	0.0016541	0.010400	1583.2	2456.9	1608.0	2612.3	3.6811	5.3142
620	15.901586	0.0017039	0.0094067	1618.6	2434.3	1645.7	2583.9	3.7396	5.2528
625	16.908567	0.0017634	0.0084538	1656.5	2407.8	1686.3	2550.7	3.8019	5.1851
630	17.969544	0.0018374	0.0075279	1697.7	2375.8	1730.7	2511.1	3.8698	5.1084
635	19.086516	0.0019353	0.0066074	1744.3	2335.7	1781.2	2461.8	3.9463	5.0181
640	20.265481	0.0020767	0.0056451	1799.7	2281.1	1841.8	2395.5	4.0375	4.9027
645	21.515425	0.0023527	0.0044553	1880.5	2185.2	1931.1	2281.0	4.1722	4.7147
647.096	22.064000	0.0031056	0.0031056	2015.7	2015.7	2084.3	2084.3	4.4070	4.4070

Table D.2   Saturation Properties of Water and Steam: Pressure Increments

P (kPa)	T (K)	v (m³/kg) sat. liquid	v (m³/kg) sat. vapor	u (kJ/kg) sat. liquid	u (kJ/kg) sat. vapor	h (kJ/kg) sat. liquid	h (kJ/kg) sat. vapor	s (kJ/kg·K) sat. liquid	s (kJ/kg·K) sat. vapor
2.0	290.64	0.0010014	66.987	73.426	2398.9	73.428	2532.9	0.26056	8.7226
4.0	302.11	0.0010041	34.791	121.38	2414.5	121.39	2553.7	0.42239	8.4734
6.0	309.31	0.0010065	23.733	151.47	2424.2	151.48	2566.6	0.52082	8.3290
8.0	314.66	0.0010085	18.099	173.83	2431.4	173.84	2576.2	0.59249	8.2273
10.0	318.96	0.0010103	14.670	191.80	2437.2	191.81	2583.9	0.6492	8.1488
12.0	322.57	0.0010119	12.358	206.90	2442.0	206.91	2590.3	0.69628	8.0849
14.0	325.70	0.0010134	10.691	219.98	2446.1	219.99	2595.8	0.73664	8.0311
16.0	328.46	0.0010147	9.4306	231.55	2449.8	231.57	2600.6	0.77201	7.9846
18.0	330.95	0.0010160	8.4431	241.95	2453.0	241.96	2605.0	0.80355	7.9437
20.0	333.21	0.0010172	7.6480	251.40	2456.0	251.42	2608.9	0.83202	7.9072
22.0	335.28	0.0010183	6.9936	260.09	2458.7	260.11	2612.5	0.85800	7.8743
24.0	337.20	0.0010193	6.4453	268.13	2461.2	268.15	2615.9	0.88191	7.8442
26.0	338.99	0.0010203	5.9792	275.62	2463.5	275.64	2619.0	0.90407	7.8167
28.0	340.67	0.0010213	5.5778	282.64	2465.7	282.66	2621.8	0.92472	7.7912
30.0	342.25	0.0010222	5.2284	289.24	2467.7	289.27	2624.5	0.94407	7.7675
32.0	343.74	0.0010231	4.9215	295.49	2469.6	295.52	2627.1	0.96228	7.7453
34.0	345.15	0.0010240	4.6497	301.41	2471.4	301.45	2629.5	0.97948	7.7246
36.0	346.50	0.0010248	4.4072	307.05	2473.1	307.09	2631.8	0.99579	7.7050
38.0	347.78	0.0010256	4.1895	312.43	2474.8	312.47	2634.0	1.0113	7.6865
40.0	349.01	0.0010264	3.9930	317.58	2476.3	317.62	2636.1	1.0261	7.6690
42.0	350.18	0.0010271	3.8146	322.52	2477.8	322.56	2638.0	1.0402	7.6524
44.0	351.32	0.0010279	3.6520	327.27	2479.3	327.31	2639.9	1.0537	7.6365
46.0	352.40	0.0010286	3.5031	331.83	2480.6	331.88	2641.8	1.0667	7.6214
48.0	353.45	0.0010293	3.3663	336.24	2481.9	336.29	2643.5	1.0792	7.6069
50.0	354.47	0.0010299	3.2400	340.49	2483.2	340.54	2645.2	1.0912	7.5930
52.0	355.45	0.0010306	3.1232	344.60	2484.4	344.66	2646.8	1.1028	7.5797
54.0	356.40	0.0010312	3.0148	348.59	2485.6	348.64	2648.4	1.1140	7.5669
56.0	357.32	0.0010319	2.9139	352.45	2486.8	352.51	2649.9	1.1248	7.5545
58.0	358.21	0.0010325	2.8198	356.20	2487.9	356.26	2651.4	1.1353	7.5426
60.0	359.08	0.0010331	2.7317	359.84	2489.0	359.91	2652.9	1.1454	7.5311
62.0	359.92	0.0010337	2.6492	363.39	2490.0	363.45	2654.2	1.1553	7.5200
64.0	360.74	0.0010342	2.5716	366.84	2491.0	366.91	2655.6	1.1649	7.5093
66.0	361.54	0.0010348	2.4986	370.20	2492.0	370.27	2656.9	1.1742	7.4989
68.0	362.32	0.0010354	2.4298	373.48	2493.0	373.55	2658.2	1.1833	7.4888
70.0	363.08	0.0010359	2.3648	376.68	2493.9	376.75	2659.4	1.1921	7.4790
72.0	363.82	0.0010364	2.3033	379.80	2494.8	379.88	2660.6	1.2007	7.4695
74.0	364.55	0.0010370	2.2450	382.86	2495.7	382.94	2661.8	1.2091	7.4602
76.0	365.26	0.0010375	2.1897	385.84	2496.5	385.92	2663.0	1.2172	7.4512
78.0	365.96	0.0010380	2.1371	388.77	2497.4	388.85	2664.1	1.2252	7.4425
80.0	366.64	0.0010385	2.0871	391.63	2498.2	391.71	2665.2	1.2330	7.4339
82.0	367.30	0.0010390	2.0394	394.43	2499.0	394.51	2666.3	1.2407	7.4256
84.0	367.95	0.0010395	1.9940	397.18	2499.8	397.26	2667.3	1.2482	7.4175
86.0	368.59	0.001040	1.9506	399.87	2500.6	399.96	2668.3	1.2555	7.4096
88.0	369.22	0.0010404	1.9091	402.51	2501.3	402.60	2669.3	1.2626	7.4018
90.0	369.84	0.0010409	1.8694	405.10	2502.1	405.20	2670.3	1.2696	7.3943
92.0	370.44	0.0010414	1.8313	407.65	2502.8	407.75	2671.3	1.2765	7.3869
94.0	371.04	0.0010418	1.7949	410.15	2503.5	410.25	2672.2	1.2833	7.3796
96.0	371.62	0.0010423	1.7599	412.61	2504.2	412.71	2673.1	1.2899	7.3726
98.0	372.19	0.0010427	1.7262	415.02	2504.9	415.13	2674.1	1.2964	7.3656
100.0	372.76	0.0010432	1.6939	417.40	2505.6	417.50	2674.9	1.3028	7.3588

P (MPa)	T (K)	v (m³/kg) sat. liquid	v (m³/kg) sat. vapor	u (kJ/kg) sat. liquid	u (kJ/kg) sat. vapor	h (kJ/kg) sat. liquid	h (kJ/kg) sat. vapor	s (kJ/kg·K) sat. liquid	s (kJ/kg·K) sat. vapor
0.10	372.76	0.0010432	1.6939	417.40	2505.6	417.50	2674.9	1.3028	7.3588
0.20	393.36	0.0010605	0.88568	504.49	2529.1	504.70	2706.2	1.5302	7.1269
0.30	406.67	0.0010732	0.60576	561.10	2543.2	561.43	2724.9	1.6717	6.9916
0.40	416.76	0.0010836	0.46238	604.22	2553.1	604.65	2738.1	1.7765	6.8955
0.50	424.98	0.0010925	0.37481	639.54	2560.7	640.09	2748.1	1.8604	6.8207
0.60	431.98	0.0011006	0.31558	669.72	2566.8	670.38	2756.1	1.9308	6.7592
0.70	438.10	0.0011080	0.27277	696.23	2571.8	697.00	2762.8	1.9918	6.7071
0.80	443.56	0.0011148	0.24034	719.97	2576.0	720.86	2768.3	2.0457	6.6616
0.90	448.50	0.0011212	0.21489	741.55	2579.6	742.56	2773.0	2.0940	6.6213
1.00	453.03	0.0011272	0.19436	761.39	2582.7	762.52	2777.1	2.1381	6.5850
1.10	457.21	0.0011330	0.17745	779.78	2585.5	781.03	2780.6	2.1785	6.5520
1.20	461.11	0.0011385	0.16326	796.96	2587.8	798.33	2783.7	2.2159	6.5217
1.30	464.75	0.0011438	0.15119	813.11	2589.9	814.60	2786.5	2.2508	6.4936
1.40	468.19	0.0011489	0.14078	828.36	2591.8	829.97	2788.8	2.2835	6.4675
1.50	471.44	0.0011539	0.13171	842.83	2593.4	844.56	2791.0	2.3143	6.4430
1.60	474.52	0.0011587	0.12374	856.60	2594.8	858.46	2792.8	2.3435	6.4199
1.70	477.46	0.0011634	0.11667	869.76	2596.1	871.74	2794.5	2.3711	6.3981
1.80	480.26	0.0011679	0.11037	882.37	2597.2	884.47	2795.9	2.3975	6.3775
1.90	482.95	0.0011724	0.10470	894.48	2598.2	896.71	2797.2	2.4227	6.3578
2.00	485.53	0.0011767	0.099585	906.14	2599.1	908.50	2798.3	2.4468	6.3390
2.10	488.01	0.0011810	0.094938	917.39	2599.9	919.87	2799.3	2.4699	6.3210
2.20	490.4	0.0011852	0.090698	928.27	2600.6	930.87	2800.1	2.4921	6.3038
2.30	492.71	0.0011894	0.086815	938.79	2601.1	941.53	2800.8	2.5136	6.2872
2.40	494.94	0.0011934	0.083244	949.00	2601.6	951.87	2801.4	2.5343	6.2712
2.50	497.10	0.0011974	0.079949	958.91	2602.1	961.91	2801.9	2.5543	6.2558
2.60	499.20	0.0012014	0.076899	968.55	2602.4	971.67	2802.3	2.5736	6.2409
2.70	501.23	0.0012053	0.074066	977.93	2602.7	981.18	2802.7	2.5924	6.2264
2.80	503.21	0.0012091	0.071429	987.07	2602.9	990.46	2802.9	2.6106	6.2124
2.90	505.13	0.0012129	0.068968	995.99	2603.1	999.51	2803.1	2.6283	6.1988
3.00	507.00	0.0012167	0.066664	1004.7	2603.2	1008.3	2803.2	2.6455	6.1856
3.10	508.83	0.0012204	0.064504	1013.2	2603.2	1017.0	2803.2	2.6623	6.1727
3.20	510.61	0.0012241	0.062475	1021.5	2603.2	1025.4	2803.1	2.6787	6.1602
3.30	512.35	0.0012278	0.060564	1029.7	2603.2	1033.7	2803.0	2.6946	6.1479
3.40	514.05	0.0012314	0.058761	1037.7	2603.1	1041.8	2802.9	2.7102	6.1360
3.50	515.71	0.0012350	0.057058	1045.5	2602.9	1049.8	2802.6	2.7254	6.1243
3.60	517.33	0.0012385	0.055446	1053.1	2602.8	1057.6	2802.4	2.7403	6.1129
3.70	518.92	0.0012421	0.053918	1060.7	2602.6	1065.3	2802.1	2.7549	6.1018
3.80	520.48	0.0012456	0.052467	1068.1	2602.3	1072.8	2801.7	2.7691	6.0908
3.90	522.01	0.0012491	0.051089	1075.3	2602.0	1080.2	2801.3	2.7831	6.0801
4.00	523.50	0.0012526	0.049776	1082.5	2601.7	1087.5	2800.8	2.7968	6.0696
4.10	524.97	0.0012560	0.048525	1089.5	2601.4	1094.7	2800.3	2.8102	6.0592
4.20	526.41	0.0012594	0.047332	1096.4	2601.0	1101.7	2799.8	2.8234	6.0491
4.30	527.83	0.0012629	0.046192	1103.2	2600.6	1108.7	2799.2	2.8363	6.0391
4.40	529.22	0.0012663	0.045102	1109.9	2600.1	1115.5	2798.6	2.8490	6.0293
4.50	530.59	0.0012696	0.044059	1116.5	2599.7	1122.2	2797.9	2.8615	6.0197
4.60	531.93	0.0012730	0.043059	1123.0	2599.2	1128.9	2797.3	2.8738	6.0102
4.70	533.25	0.0012764	0.042100	1129.5	2598.7	1135.5	2796.5	2.8859	6.0009
4.80	534.55	0.0012797	0.041180	1135.8	2598.1	1141.9	2795.8	2.8978	5.9917
4.90	535.83	0.0012831	0.040296	1142.0	2597.6	1148.3	2795.0	2.9095	5.9826
5.00	537.09	0.0012864	0.039446	1148.2	2597.0	1154.6	2794.2	2.9210	5.9737

(continued)

Table D.2   (continued)

P (MPa)	T (K)	v (m³/kg) sat. liquid	v (m³/kg) sat. vapor	u (kJ/kg) sat. liquid	u (kJ/kg) sat. vapor	h (kJ/kg) sat. liquid	h (kJ/kg) sat. vapor	s (kJ/kg·K) sat. liquid	s (kJ/kg·K) sat. vapor
5.0	537.09	0.0012864	0.039446	1148.2	2597.0	1154.6	2794.2	2.9210	5.9737
5.5	543.12	0.0013029	0.035642	1177.9	2593.7	1185.1	2789.7	2.9762	5.9307
6.0	548.73	0.0013193	0.032448	1206.0	2589.9	1213.9	2784.6	3.0278	5.8901
6.5	554.01	0.0013356	0.029727	1232.7	2585.7	1241.4	2778.9	3.0764	5.8516
7.0	558.98	0.0013519	0.027378	1258.2	2581	1267.7	2772.6	3.1224	5.8148
7.5	563.69	0.0013682	0.025330	1282.7	2575.9	1292.9	2765.9	3.1662	5.7793
8.0	568.16	0.0013847	0.023526	1306.2	2570.5	1317.3	2758.7	3.2081	5.7450
8.5	572.42	0.0014013	0.021923	1329.0	2564.7	1340.9	2751.0	3.2483	5.7117
9.0	576.49	0.0014181	0.020490	1351.1	2558.5	1363.9	2742.9	3.2870	5.6791
9.5	580.40	0.0014352	0.019199	1372.6	2552.0	1386.2	2734.4	3.3244	5.6473
10.0	584.15	0.0014526	0.018030	1393.5	2545.2	1408.1	2725.5	3.3606	5.6160
10.5	587.75	0.0014703	0.016965	1414.0	2538.0	1429.4	2716.1	3.3959	5.5851
11.0	591.23	0.0014885	0.015990	1434.1	2530.5	1450.4	2706.3	3.4303	5.5545
11.5	594.58	0.0015071	0.015093	1453.8	2522.6	1471.1	2696.1	3.4638	5.5241
12.0	597.83	0.0015263	0.014264	1473.1	2514.3	1491.5	2685.4	3.4967	5.4939
12.5	600.96	0.0015461	0.013496	1492.3	2505.6	1511.6	2674.3	3.5290	5.4638
13.0	604.00	0.0015665	0.012780	1511.1	2496.5	1531.5	2662.7	3.5608	5.4336
13.5	606.95	0.0015877	0.012112	1529.9	2487.0	1551.3	2650.5	3.5921	5.4032
14.0	609.82	0.0016097	0.011485	1548.4	2477.1	1571.0	2637.9	3.6232	5.3727
14.5	612.60	0.0016328	0.010895	1566.9	2466.6	1590.6	2624.6	3.6539	5.3418
15.0	615.31	0.0016570	0.010338	1585.3	2455.6	1610.2	2610.7	3.6846	5.3106
15.5	617.94	0.0016824	0.0098106	1603.8	2444.1	1629.9	2596.1	3.7151	5.2788
16.0	620.50	0.0017094	0.0093088	1622.3	2431.8	1649.7	2580.8	3.7457	5.2463
16.5	623.00	0.0017383	0.0088299	1641.0	2418.9	1669.7	2564.6	3.7765	5.2130
17.0	625.44	0.0017693	0.0083709	1659.9	2405.2	1690.0	2547.5	3.8077	5.1787
17.5	627.82	0.0018029	0.0079292	1679.2	2390.5	1710.8	2529.3	3.8394	5.1431
18.0	630.14	0.0018398	0.0075017	1699.0	2374.8	1732.1	2509.8	3.8718	5.1061
18.5	632.41	0.0018807	0.0070856	1719.3	2357.8	1754.1	2488.8	3.9053	5.0670
19.0	634.62	0.0019268	0.0066773	1740.5	2339.1	1777.2	2466.0	3.9401	5.0256
19.5	636.79	0.0019792	0.0062725	1762.8	2318.5	1801.4	2440.8	3.9767	4.9808
20.0	638.90	0.002040	0.0058652	1786.4	2295.0	1827.2	2412.3	4.0156	4.9314
20.5	640.96	0.0021126	0.0054457	1812.0	2267.6	1855.3	2379.2	4.0579	4.8753
21.0	642.98	0.0022055	0.0049961	1841.2	2233.7	1887.6	2338.6	4.1064	4.8079
21.5	644.94	0.0023468	0.0044734	1879.1	2186.9	1929.5	2283.1	4.1698	4.7181
22.0	646.86	0.0027044	0.0036475	1951.8	2092.8	2011.3	2173.1	4.2945	4.5446
22.064	647.096	0.0031056	0.0031056	2015.7	2015.7	2084.3	2084.3	4.4070	4.4070

Table D.3 Superheated Vapor (Steam)*

Table D.3A Isobaric Data for $P = 0.006$ MPa

T (K)	P (MPa)	ρ (kg/m³)	v (m³/kg)	u (kJ/kg)	h (kJ/kg)	s (kJ/kg·K)
309.31	0.006	0.042135	23.733	2424.2	2566.6	8.3290
320	0.006	0.040708	24.565	2439.7	2587.1	8.3940
340	0.006	0.038291	26.116	2468.4	2625.1	8.5092
360	0.006	0.036151	27.662	2497.0	2663.0	8.6176
380	0.006	0.034239	29.206	2525.7	2701.0	8.7202
400	0.006	0.032522	30.748	2554.6	2739.0	8.8179
420	0.006	0.030969	32.290	2583.5	2777.3	8.9112
440	0.006	0.029559	33.831	2612.7	2815.7	9.0005
460	0.006	0.028272	35.371	2642.1	2854.3	9.0863
480	0.006	0.027092	36.911	2671.6	2893.1	9.1689
500	0.006	0.026008	38.450	2701.4	2932.1	9.2485
520	0.006	0.025006	39.990	2731.4	2971.4	9.3255
540	0.006	0.024079	41.529	2761.7	3010.9	9.4000
560	0.006	0.023219	43.068	2792.2	3050.6	9.4722
580	0.006	0.022418	44.607	2822.9	3090.6	9.5424
600	0.006	0.02167	46.146	2853.9	3130.8	9.6105
620	0.006	0.020971	47.685	2885.1	3171.2	9.6769
640	0.006	0.020315	49.224	2916.6	3211.9	9.7415
660	0.006	0.019699	50.763	2948.3	3252.9	9.8046
680	0.006	0.01912	52.301	2980.4	3294.2	9.8661
700	0.006	0.018573	53.840	3012.6	3335.7	9.9263
720	0.006	0.018057	55.379	3045.2	3377.4	9.9851
740	0.006	0.017569	56.917	3078.0	3419.5	10.043
760	0.006	0.017107	58.456	3111.0	3461.8	10.099
780	0.006	0.016668	59.995	3144.4	3504.3	10.154
800	0.006	0.016251	61.533	3178.0	3547.2	10.209

*Property values generated from NIST Database 23: REFPROP Version 7.0 (August 2002).

Table D.3B Isobaric Data for $P = 0.035$ MPa

T (K)	P (MPa)	ρ (kg/m³)	v (m³/kg)	u (kJ/kg)	h (kJ/kg)	s (kJ/kg·K)
345.83	0.035	0.22099	4.5251	2472.3	2630.7	7.7146
360	0.035	0.21197	4.7176	2493.5	2658.6	7.7939
380	0.035	0.20052	4.9871	2523.1	2697.7	7.8994
400	0.035	0.1903	5.2549	2552.5	2736.5	7.9989
420	0.035	0.1811	5.5218	2581.9	2775.2	8.0934
440	0.035	0.17278	5.7879	2611.4	2813.9	8.1835
460	0.035	0.16519	6.0535	2640.9	2852.8	8.2699
480	0.035	0.15826	6.3187	2670.7	2891.8	8.353
500	0.035	0.15189	6.5837	2700.6	2931.0	8.433
520	0.035	0.14602	6.8484	2730.7	2970.4	8.5102
540	0.035	0.14059	7.113	2761.1	3010.0	8.5849
560	0.035	0.13555	7.3775	2791.6	3049.8	8.6573
580	0.035	0.13086	7.6419	2822.4	3089.9	8.7276
600	0.035	0.12648	7.9062	2853.4	3130.1	8.7958
620	0.035	0.12239	8.1704	2884.7	3170.7	8.8623
640	0.035	0.11856	8.4346	2916.2	3211.4	8.927
660	0.035	0.11496	8.6987	2948.0	3252.5	8.9901
680	0.035	0.11157	8.9627	2980.0	3293.7	9.0517
700	0.035	0.10838	9.2268	3012.3	3335.3	9.1119
720	0.035	0.10537	9.4908	3044.9	3377.1	9.1708
740	0.035	0.10251	9.7547	3077.7	3419.1	9.2284
760	0.035	0.099813	10.019	3110.8	3461.4	9.2848
780	0.035	0.097251	10.283	3144.2	3504.0	9.3402
800	0.035	0.094818	10.547	3177.8	3546.9	9.3944
820	0.035	0.092503	10.81	3211.7	3590.1	9.4477
840	0.035	0.090299	11.074	3245.9	3633.5	9.5

Table D.3C   Isobaric Data for P = 0.070 MPa

T (K)	P (MPa)	ρ (kg/m³)	v (m³/kg)	u (kJ/kg)	h (kJ/kg)	s (kJ/kg·K)
363.08	0.07	0.42287	2.3648	2493.9	2659.4	7.479
380	0.07	0.40301	2.4813	2519.9	2693.5	7.5709
400	0.07	0.38205	2.6175	2550.0	2733.2	7.6727
420	0.07	0.3633	2.7525	2579.9	2772.6	7.7687
440	0.07	0.3464	2.8868	2609.7	2811.8	7.8599
460	0.07	0.33106	3.0206	2639.6	2851.0	7.9471
480	0.07	0.31706	3.154	2669.5	2890.3	8.0307
500	0.07	0.30422	3.2871	2699.6	2929.7	8.1111
520	0.07	0.2924	3.42	2729.9	2969.3	8.1886
540	0.07	0.28147	3.5528	2760.3	3009.0	8.2636
560	0.07	0.27134	3.6854	2790.9	3048.9	8.3362
580	0.07	0.26193	3.8179	2821.8	3089.0	8.4066
600	0.07	0.25314	3.9503	2852.9	3129.4	8.475
620	0.07	0.24494	4.0827	2884.2	3170.0	8.5416
640	0.07	0.23725	4.215	2915.8	3210.8	8.6064
660	0.07	0.23003	4.3472	2947.6	3251.9	8.6696
680	0.07	0.22324	4.4794	2979.6	3293.2	8.7312
700	0.07	0.21685	4.6116	3012.0	3334.8	8.7915
720	0.07	0.2108	4.7437	3044.5	3376.6	8.8504
740	0.07	0.20509	4.8758	3077.4	3418.7	8.9081
760	0.07	0.19968	5.0079	3110.5	3461.1	8.9645
780	0.07	0.19455	5.14	3143.9	3503.7	9.0199
800	0.07	0.18968	5.2721	3177.5	3546.6	9.0742
820	0.07	0.18504	5.4041	3211.5	3589.7	9.1275
840	0.07	0.18063	5.5361	3245.7	3633.2	9.1798
860	0.07	0.17643	5.6681	3280.1	3676.9	9.2313

Table D.3D   Isobaric Data for P = 0.100 MPa

T (K)	P (MPa)	ρ (kg/m³)	v (m³/kg)	u (kJ/kg)	h (kJ/kg)	s (kJ/kg·K)
372.76	0.1	0.59034	1.6939	2505.6	2674.9	7.3588
380	0.1	0.57824	1.7294	2517.0	2689.9	7.3986
400	0.1	0.54761	1.8261	2547.8	2730.4	7.5025
420	0.1	0.52038	1.9217	2578.2	2770.3	7.5999
440	0.1	0.49592	2.0165	2608.3	2810.0	7.6921
460	0.1	0.47378	2.1107	2638.4	2849.5	7.7799
480	0.1	0.45361	2.2045	2668.5	2889.0	7.864
500	0.1	0.43514	2.2981	2698.7	2928.6	7.9447
520	0.1	0.41815	2.3915	2729.1	2968.2	8.0226
540	0.1	0.40247	2.4847	2759.6	3008.1	8.0978
560	0.1	0.38794	2.5777	2790.3	3048.1	8.1705
580	0.1	0.37444	2.6707	2821.3	3088.3	8.2411
600	0.1	0.36185	2.7635	2852.4	3128.8	8.3096
620	0.1	0.3501	2.8563	2883.8	3169.4	8.3763
640	0.1	0.33909	2.9491	2915.4	3210.3	8.4411
660	0.1	0.32876	3.0418	2947.2	3251.4	8.5044
680	0.1	0.31904	3.1344	2979.3	3292.8	8.5661
700	0.1	0.30988	3.227	3011.7	3334.4	8.6264
720	0.1	0.30124	3.3196	3044.3	3376.2	8.6854
740	0.1	0.29307	3.4122	3077.1	3418.3	8.7431
760	0.1	0.28533	3.5047	3110.3	3460.7	8.7996
780	0.1	0.27799	3.5972	3143.6	3503.4	8.855
800	0.1	0.27102	3.6897	3177.3	3546.3	8.9093
820	0.1	0.2644	3.7822	3211.3	3589.5	8.9626
840	0.1	0.25809	3.8747	3245.5	3632.9	9.015
860	0.1	0.25207	3.9671	3280.0	3676.7	9.0665
880	0.1	0.24633	4.0595	3314.7	3720.7	9.1171

Table D.3E   Isobaric Data for $P = 0.150$ MPa

T (K)	P (MPa)	ρ (kg/m³)	v (m³/kg)	u (kJ/kg)	h (kJ/kg)	s (kJ/kg·K)
384.5	0.15	0.8626	1.1593	2519.2	2693.1	7.223
400	0.15	0.82612	1.2105	2544.0	2725.6	7.3058
420	0.15	0.78408	1.2754	2575.2	2766.5	7.4057
440	0.15	0.74658	1.3394	2605.9	2806.8	7.4995
460	0.15	0.71278	1.403	2636.4	2846.9	7.5884
480	0.15	0.6821	1.4661	2666.9	2886.8	7.6733
500	0.15	0.65407	1.5289	2697.3	2926.6	7.7547
520	0.15	0.62835	1.5915	2727.8	2966.6	7.833
540	0.15	0.60463	1.6539	2758.5	3006.6	7.9086
560	0.15	0.58268	1.7162	2789.4	3046.8	7.9816
580	0.15	0.5623	1.7784	2820.4	3087.1	8.0524
600	0.15	0.54333	1.8405	2851.6	3127.7	8.1212
620	0.15	0.52562	1.9025	2883.1	3168.4	8.188
640	0.15	0.50903	1.9645	2914.7	3209.4	8.253
660	0.15	0.49348	2.0264	2946.6	3250.6	8.3164
680	0.15	0.47886	2.0883	2978.8	3292.0	8.3782
700	0.15	0.46508	2.1502	3011.1	3333.7	8.4386
720	0.15	0.45209	2.212	3043.8	3375.6	8.4976
740	0.15	0.4398	2.2738	3076.7	3417.7	8.5554
760	0.15	0.42817	2.3355	3109.8	3460.2	8.6119
780	0.15	0.41714	2.3973	3143.3	3502.9	8.6674
800	0.15	0.40667	2.459	3177.0	3545.8	8.7217
820	0.15	0.39671	2.5207	3210.9	3589.0	8.7751
840	0.15	0.38724	2.5824	3245.1	3632.5	8.8275
860	0.15	0.3782	2.6441	3279.7	3676.3	8.879
880	0.15	0.36958	2.7058	3314.4	3720.3	8.9296
900	0.15	0.36135	2.7674	3349.5	3764.6	8.9794

Table D.3F   Isobaric Data for $P = 0.300$ MPa

T (K)	P (MPa)	ρ (kg/m³)	v (m³/kg)	u (kJ/kg)	h (kJ/kg)	s (kJ/kg·K)
406.67	0.3	1.6508	0.60576	2543.2	2724.9	6.9916
420	0.3	1.5906	0.62868	2565.8	2754.4	7.063
440	0.3	1.5101	0.6622	2598.4	2797.1	7.1623
460	0.3	1.4387	0.69507	2630.3	2838.8	7.255
480	0.3	1.3746	0.72749	2661.7	2879.9	7.3426
500	0.3	1.3165	0.75958	2692.9	2920.8	7.426
520	0.3	1.2635	0.79144	2724.0	2961.5	7.5057
540	0.3	1.2149	0.82312	2755.2	3002.1	7.5824
560	0.3	1.17	0.85467	2786.4	3042.8	7.6564
580	0.3	1.1285	0.8861	2817.7	3083.6	7.7279
600	0.3	1.09	0.91744	2849.2	3124.4	7.7973
620	0.3	1.0541	0.94871	2880.9	3165.5	7.8646
640	0.3	1.0205	0.97992	2912.7	3206.7	7.93
660	0.3	0.98904	1.0111	2944.8	3248.1	7.9937
680	0.3	0.95951	1.0422	2977.1	3289.7	8.0558
700	0.3	0.93173	1.0733	3009.6	3331.6	8.1165
720	0.3	0.90553	1.1043	3042.4	3373.6	8.1757
740	0.3	0.88079	1.1353	3075.3	3416.0	8.2337
760	0.3	0.85738	1.1663	3108.6	3458.5	8.2904
780	0.3	0.8352	1.1973	3142.1	3501.3	8.346
800	0.3	0.81415	1.2283	3175.9	3544.3	8.4005
820	0.3	0.79414	1.2592	3209.9	3587.6	8.4539
840	0.3	0.7751	1.2901	3244.2	3631.2	8.5064
860	0.3	0.75696	1.3211	3278.7	3675.1	8.558
880	0.3	0.73966	1.352	3313.6	3719.2	8.6087
900	0.3	0.72314	1.3829	3348.7	3763.5	8.6586
920	0.3	0.70734	1.4138	3384.1	3808.2	8.7076

Table D.3G   Isobaric Data for $P = 0.50$ MPa

T (K)	P (MPa)	ρ (kg/m³)	v (m³/kg)	u (kJ/kg)	h (kJ/kg)	s (kJ/kg·K)
424.98	0.5	2.668	0.37481	2560.7	2748.1	6.8207
440	0.5	2.5579	0.39095	2587.6	2783.1	6.9015
460	0.5	2.429	0.41169	2621.6	2827.4	7.0001
480	0.5	2.3153	0.43191	2654.5	2870.5	7.0917
500	0.5	2.2135	0.45176	2686.8	2912.7	7.1779
520	0.5	2.1215	0.47136	2718.8	2954.5	7.2599
540	0.5	2.0376	0.491	2750.6	2996.0	7.3382
560	0.5	1.9607	0.510	2782.4	3037.4	7.4134
580	0.5	1.8898	0.529	2814.1	3078.7	7.486
600	0.5	1.8242	0.548	2846.0	3120.1	7.5561
620	0.5	1.7631	0.567	2878.0	3161.5	7.6241
640	0.5	1.7063	0.586	2910.1	3203.1	7.6901
660	0.5	1.6531	0.605	2942.4	3244.8	7.7542
680	0.5	1.6032	0.624	2974.8	3286.7	7.8168
700	0.5	1.5564	0.643	3007.5	3328.8	7.8777
720	0.5	1.5123	0.661	3040.4	3371.1	7.9373
740	0.5	1.4706	0.680	3073.6	3413.5	7.9955
760	0.5	1.4313	0.699	3106.9	3456.3	8.0524
780	0.5	1.394	0.717	3140.5	3499.2	8.1082
800	0.5	1.3587	0.736	3174.4	3542.4	8.1629
820	0.5	1.3252	0.755	3208.5	3585.8	8.2165
840	0.5	1.2932	0.77325	3242.9	3629.5	8.2691
860	0.5	1.2629	0.79186	3277.5	3673.4	8.3208
880	0.5	1.2339	0.81045	3312.4	3717.6	8.3716
900	0.5	1.2062	0.82904	3347.6	3762.1	8.4216
920	0.5	1.1798	0.84762	3383.0	3806.8	8.4708
940	0.5	1.1545	0.86619	3418.7	3851.8	8.5191

Table D.3H   Isobaric Data for $P = 0.70$ MPa

T (K)	P (MPa)	ρ (kg/m³)	v (m³/kg)	u (kJ/kg)	h (kJ/kg)	s (kJ/kg·K)
438.1	0.7	3.666	0.27277	2571.8	2762.8	6.7071
440	0.7	3.6453	0.27433	2575.5	2767.6	6.718
460	0.7	3.4477	0.29005	2612.3	2815.3	6.8242
480	0.7	3.2775	0.30511	2647.0	2860.5	6.9204
500	0.7	3.1274	0.31976	2680.5	2904.3	7.0099
520	0.7	2.9929	0.33413	2713.4	2947.3	7.0941
540	0.7	2.8712	0.34828	2745.9	2989.7	7.1742
560	0.7	2.7603	0.36228	2778.3	3031.9	7.2508
580	0.7	2.6585	0.37616	2810.5	3073.8	7.3244
600	0.7	2.5645	0.38994	2842.7	3115.7	7.3953
620	0.7	2.4775	0.40364	2875.0	3157.6	7.464
640	0.7	2.3965	0.41728	2907.4	3199.5	7.5306
660	0.7	2.3209	0.43086	2939.9	3241.5	7.5952
680	0.7	2.2502	0.4444	2972.6	3283.7	7.6582
700	0.7	2.1839	0.4579	3005.4	3326.0	7.7195
720	0.7	2.1215	0.47137	3038.5	3368.5	7.7793
740	0.7	2.0626	0.48481	3071.8	3411.1	7.8378
760	0.7	2.0071	0.49823	3105.3	3454.0	7.895
780	0.7	1.9545	0.51163	3139.0	3497.1	7.9509
800	0.7	1.9047	0.52501	3172.9	3540.4	8.0058
820	0.7	1.8575	0.53837	3207.1	3584.0	8.0595
840	0.7	1.8125	0.55172	3241.6	3627.8	8.1123
860	0.7	1.7697	0.56505	3276.3	3671.8	8.1641
880	0.7	1.729	0.57838	3311.3	3716.1	8.215
900	0.7	1.6901	0.59169	3346.5	3760.7	8.2651
920	0.7	1.6529	0.60499	3382.0	3805.5	8.3143
940	0.7	1.6174	0.61829	3417.7	3850.5	8.3628

Table D.3I Isobaric Data for $P = 1.0$ MPa

T (K)	P (MPa)	ρ (kg/m³)	v (m³/kg)	u (kJ/kg)	h (kJ/kg)	s (kJ/kg·K)
453.03	1	5.145	0.19436	2582.7	2777.1	6.585
460	1	5.0376	0.19851	2597.0	2795.5	6.6253
480	1	4.7658	0.20983	2634.9	2844.7	6.7301
500	1	4.5323	0.22064	2670.6	2891.2	6.825
520	1	4.3268	0.23112	2705.0	2936.1	6.9131
540	1	4.1431	0.24137	2738.7	2980.1	6.996
560	1	3.9771	0.25144	2772.0	3023.4	7.0748
580	1	3.8258	0.26138	2804.9	3066.3	7.1501
600	1	3.6871	0.27122	2837.7	3109.0	7.2224
620	1	3.5591	0.28097	2870.5	3151.5	7.2921
640	1	3.4404	0.29066	2903.3	3194.0	7.3596
660	1	3.33	0.3003	2936.2	3236.5	7.425
680	1	3.227	0.30988	2969.2	3279.1	7.4885
700	1	3.1305	0.31943	3002.3	3321.7	7.5504
720	1	3.04	0.32895	3035.6	3364.5	7.6107
740	1	2.9547	0.33844	3069.1	3407.5	7.6695
760	1	2.8744	0.3479	3102.7	3450.6	7.727
780	1	2.7984	0.35734	3136.6	3494.0	7.7833
800	1	2.7265	0.36677	3170.7	3537.5	7.8384
820	1	2.6583	0.37618	3205.1	3581.2	7.8924
840	1	2.5936	0.38557	3239.6	3625.2	7.9454
860	1	2.532	0.39495	3274.4	3669.4	7.9974
880	1	2.4733	0.40432	3309.5	3713.8	8.0484
900	1	2.4174	0.41367	3344.8	3758.5	8.0986
920	1	2.3639	0.42302	3380.4	3803.4	8.148
940	1	2.3129	0.43236	3416.3	3848.6	8.1966
960	1	2.264	0.4417	3452.4	3894.1	8.2444

Table D.3J Isobaric Data for $P = 1.5$ MPa

T (K)	P (MPa)	ρ (kg/m³)	v (m³/kg)	u (kJ/kg)	h (kJ/kg)	s (kJ/kg·K)
471.44	1.5	7.5924	0.13171	2593.4	2791.0	6.443
480	1.5	7.3885	0.13534	2612.3	2815.3	6.4942
500	1.5	6.9775	0.14332	2652.6	2867.6	6.6009
520	1.5	6.629	0.15085	2690.1	2916.4	6.6967
540	1.5	6.3249	0.1581	2726.1	2963.2	6.7851
560	1.5	6.0549	0.16515	2761.0	3008.8	6.8678
580	1.5	5.8121	0.17206	2795.3	3053.4	6.9462
600	1.5	5.5915	0.17884	2829.2	3097.5	7.0209
620	1.5	5.3897	0.18554	2862.9	3141.2	7.0925
640	1.5	5.2039	0.19216	2896.4	3184.7	7.1616
660	1.5	5.0319	0.19873	2929.9	3228.0	7.2282
680	1.5	4.8721	0.20525	2963.4	3271.3	7.2929
700	1.5	4.7231	0.21173	2997.0	3314.6	7.3556
720	1.5	4.5836	0.21817	3030.7	3358.0	7.4167
740	1.5	4.4527	0.22458	3064.5	3401.4	7.4762
760	1.5	4.3295	0.23097	3098.5	3445.0	7.5343
780	1.5	4.2133	0.23734	3132.7	3488.7	7.5911
800	1.5	4.1036	0.24369	3167.0	3532.6	7.6466
820	1.5	3.9997	0.25002	3201.6	3576.6	7.701
840	1.5	3.9011	0.25634	3236.4	3620.9	7.7543
860	1.5	3.8075	0.26264	3271.4	3665.3	7.8066
880	1.5	3.7184	0.26893	3306.6	3710.0	7.858
900	1.5	3.6335	0.27522	3342.1	3754.9	7.9084
920	1.5	3.5525	0.28149	3377.8	3800.0	7.958
940	1.5	3.4752	0.28775	3413.8	3845.4	8.0068
960	1.5	3.4013	0.29401	3450.0	3891.0	8.0548
980	1.5	3.3305	0.30026	3486.5	3936.9	8.1021

Table D.3K   Isobaric Data for $P = 2.0$ MPa

T (K)	P (MPa)	ρ (kg/m³)	v (m³/kg)	u (kJ/kg)	h (kJ/kg)	s (kJ/kg·K)
485.53	2	10.042	0.099585	2599.1	2798.3	6.339
500	2	9.5781	0.10441	2632.6	2841.4	6.4265
520	2	9.0447	0.11056	2674.0	2895.1	6.5319
540	2	8.5934	0.11637	2712.6	2945.4	6.6267
560	2	8.2008	0.12194	2749.5	2993.4	6.7141
580	2	7.853	0.12734	2785.3	3040.0	6.7959
600	2	7.5406	0.13262	2820.4	3085.6	6.8732
620	2	7.2573	0.13779	2855.0	3130.6	6.947
640	2	6.9983	0.14289	2889.4	3175.1	7.0176
660	2	6.7599	0.14793	2923.5	3219.4	7.0857
680	2	6.5395	0.15292	2957.6	3263.4	7.1514
700	2	6.3346	0.15786	2991.6	3307.4	7.2151
720	2	6.1436	0.16277	3025.8	3351.3	7.277
740	2	5.9648	0.16765	3059.9	3395.2	7.3372
760	2	5.7969	0.1725	3094.2	3439.3	7.3959
780	2	5.639	0.17734	3128.7	3483.4	7.4532
800	2	5.4901	0.18215	3163.3	3527.6	7.5092
820	2	5.3493	0.18694	3198.1	3572.0	7.564
840	2	5.2159	0.19172	3233.1	3616.5	7.6176
860	2	5.0894	0.19649	3268.3	3661.2	7.6702
880	2	4.9691	0.20124	3303.7	3706.1	7.7219
900	2	4.8547	0.20599	3339.3	3751.3	7.7726
920	2	4.7456	0.21072	3375.2	3796.6	7.8224
940	2	4.6415	0.21545	3411.3	3842.2	7.8714
960	2	4.5421	0.22016	3447.6	3887.9	7.9196
980	2	4.4469	0.22488	3484.2	3934.0	7.967
1000	2	4.3558	0.22958	3521.1	3980.2	8.0137

Table D.3L   Isobaric Data for $P = 3.0$ MPa

T (K)	P (MPa)	ρ (kg/m³)	v (m³/kg)	u (kJ/kg)	h (kJ/kg)	s (kJ/kg·K)
507	3	15.001	0.066664	2603.2	2803.2	6.1856
520	3	14.309	0.069888	2637.1	2846.7	6.2704
540	3	13.442	0.074391	2682.9	2906.1	6.3825
560	3	12.729	0.078561	2724.7	2960.4	6.4812
580	3	12.119	0.082512	2764.1	3011.6	6.5712
600	3	11.587	0.086307	2801.9	3060.8	6.6546
620	3	11.113	0.089986	2838.7	3108.6	6.733
640	3	10.687	0.093576	2874.7	3155.5	6.8073
660	3	10.299	0.097096	2910.3	3201.6	6.8783
680	3	9.9443	0.10056	2945.6	3247.3	6.9465
700	3	9.6174	0.10398	2980.7	3292.6	7.0122
720	3	9.3147	0.10736	3015.7	3337.7	7.0758
740	3	9.033	0.1107	3050.6	3382.7	7.1374
760	3	8.77	0.11403	3085.6	3427.7	7.1973
780	3	8.5236	0.11732	3120.6	3472.6	7.2557
800	3	8.292	0.1206	3155.8	3517.6	7.3126
820	3	8.0738	0.12386	3191.0	3562.6	7.3682
840	3	7.8678	0.1271	3226.4	3607.7	7.4226
860	3	7.6729	0.13033	3262.0	3653.0	7.4758
880	3	7.488	0.13355	3297.8	3698.4	7.528
900	3	7.3124	0.13675	3333.7	3744.0	7.5792
920	3	7.1454	0.13995	3369.9	3789.7	7.6295
940	3	6.9863	0.14314	3406.2	3835.7	7.6789
960	3	6.8344	0.14632	3442.8	3881.8	7.7275
980	3	6.6893	0.14949	3479.7	3928.1	7.7752
1000	3	6.5506	0.15266	3516.7	3974.7	7.8223
1020	3	6.4177	0.15582	3554.0	4021.5	7.8686

Table D.3M **Isobaric Data for P = 4.0 MPa**

T (K)	P (MPa)	ρ (kg/m^3)	v (m^3/kg)	u (kJ/kg)	h (kJ/kg)	s (kJ/kg$\cdot$K)
523.5	4	20.09	0.049776	2601.7	2800.8	6.0696
540	4	18.831	0.053103	2648.4	2860.8	6.1824
560	4	17.642	0.056683	2696.9	2923.7	6.2968
580	4	16.675	0.05997	2740.9	2980.8	6.397
600	4	15.857	0.063063	2782.1	3034.3	6.4878
620	4	15.147	0.066018	2821.4	3085.4	6.5716
640	4	14.52	0.069	2859.4	3134.9	6.6501
660	4	13.958	0.072	2896.6	3183.2	6.7244
680	4	13.449	0.074	2933.2	3230.6	6.7953
700	4	12.985	0.077	2969.4	3277.5	6.8632
720	4	12.558	0.080	3005.4	3323.9	6.9285
740	4	12.163	0.082	3041.1	3370.0	6.9917
760	4	11.796	0.085	3076.8	3415.9	7.0529
780	4	11.454	0.087	3112.5	3461.7	7.1124
800	4	11.134	0.090	3148.2	3507.4	7.1702
820	4	10.833	0.092	3183.9	3553.1	7.2267
840	4	10.55	0.095	3219.7	3598.9	7.2818
860	4	10.283	0.097	3255.7	3644.7	7.3357
880	4	10.03	0.100	3291.8	3690.6	7.3885
900	4	9.791	0.102	3328.1	3736.6	7.4402
920	4	9.5636	0.105	3364.5	3782.8	7.4909
940	4	9.3473	0.10698	3401.2	3829.1	7.5407
960	4	9.1412	0.10939	3438.0	3875.6	7.5897
980	4	8.9446	0.1118	3475.1	3922.3	7.6378
1000	4	8.7568	0.1142	3512.4	3969.1	7.6851
1020	4	8.5771	0.11659	3549.9	4016.2	7.7317
1040	4	8.405	0.11898	3587.6	4063.5	7.7776

Table D.3N **Isobaric Data for P = 6.0 MPa**

T (K)	P (MPa)	ρ (kg/m^3)	v (m^3/kg)	u (kJ/kg)	h (kJ/kg)	s (kJ/kg$\cdot$K)
548.73	6	30.818	0.032448	2589.9	2784.6	5.8901
560	6	29.166	0.034287	2629.1	2834.8	5.9807
580	6	26.947	0.03711	2687.2	2909.8	6.1125
600	6	25.247	0.039608	2737.6	2975.2	6.2233
620	6	23.865	0.041903	2783.5	3034.9	6.3211
640	6	22.697	0.044058	2826.5	3090.8	6.4099
660	6	21.686	0.046112	2867.5	3144.2	6.4921
680	6	20.795	0.048089	2907.2	3195.7	6.569
700	6	19.998	0.050006	2945.9	3246.0	6.6418
720	6	19.277	0.051875	2984.0	3295.2	6.7112
740	6	18.62	0.053706	3021.5	3343.8	6.7777
760	6	18.017	0.055504	3058.7	3391.8	6.8417
780	6	17.46	0.057274	3095.7	3439.4	6.9036
800	6	16.943	0.059022	3132.6	3486.7	6.9635
820	6	16.461	0.06075	3169.4	3533.9	7.0217
840	6	16.01	0.062461	3206.1	3580.9	7.0784
860	6	15.587	0.064157	3242.9	3627.9	7.1336
880	6	15.188	0.06584	3279.8	3674.8	7.1876
900	6	14.812	0.067512	3316.7	3721.8	7.2404
920	6	14.457	0.069173	3353.8	3768.8	7.2921
940	6	14.119	0.070825	3391.0	3815.9	7.3427
960	6	13.799	0.072469	3428.4	3863.2	7.3924
980	6	13.494	0.074105	3465.9	3910.5	7.4413
1000	6	13.204	0.075735	3503.6	3958.0	7.4892
1020	6	12.927	0.077358	3541.5	4005.6	7.5364
1040	6	12.662	0.078977	3579.6	4053.4	7.5828
1060	6	12.409	0.08059	3617.9	4101.4	7.6285

Table D.3O   Isobaric Data for $P = 8.0$ MPa

T (K)	P (MPa)	ρ (kg/m³)	v (m³/kg)	u (kJ/kg)	h (kJ/kg)	s (kJ/kg·K)
568.16	8	42.507	0.023526	2570.5	2758.7	5.745
580	8	39.647	0.025223	2619.0	2820.7	5.8531
600	8	36.218	0.027611	2684.7	2905.6	5.9971
620	8	33.704	0.02967	2740.2	2977.6	6.1151
640	8	31.713	0.031532	2789.9	3042.1	6.2176
660	8	30.065	0.033261	2835.8	3101.9	6.3096
680	8	28.658	0.034894	2879.3	3158.4	6.394
700	8	27.431	0.036455	2921.0	3212.6	6.4726
720	8	26.343	0.037961	2961.5	3265.2	6.5466
740	8	25.367	0.039422	3001.1	3316.5	6.6168
760	8	24.482	0.040847	3040.0	3366.8	6.6839
780	8	23.673	0.042242	3078.5	3416.4	6.7484
800	8	22.929	0.043612	3116.6	3465.5	6.8105
820	8	22.242	0.044961	3154.5	3514.2	6.8706
840	8	21.602	0.046292	3192.3	3562.6	6.9289
860	8	21.006	0.047606	3229.9	3610.8	6.9856
880	8	20.447	0.048907	3267.6	3658.8	7.0409
900	8	19.922	0.050196	3305.2	3706.8	7.0948
920	8	19.427	0.051475	3342.9	3754.7	7.1474
940	8	18.96	0.052744	3380.7	3802.6	7.199
960	8	18.517	0.054004	3418.6	3850.6	7.2495
980	8	18.097	0.055257	3456.6	3898.6	7.299
1000	8	17.698	0.056503	3494.7	3946.8	7.3476
1020	8	17.318	0.057742	3533.1	3995.0	7.3953
1040	8	16.956	0.058976	3571.5	4043.3	7.4423
1060	8	16.61	0.060205	3610.2	4091.8	7.4885
1080	8	16.279	0.06143	3649.0	4140.5	7.5339

Table D.3P   Isobaric Data for $P = 10.0$ MPa

T (K)	P (MPa)	ρ (kg/m³)	v (m³/kg)	u (kJ/kg)	h (kJ/kg)	s (kJ/kg·K)
584.15	10	55.463	0.01803	2545.2	2725.5	5.616
600	10	49.773	0.020091	2619.1	2820.0	5.7756
620	10	45.151	0.022148	2689.7	2911.2	5.9253
640	10	41.84	0.0239	2748.7	2987.7	6.0468
660	10	39.259	0.025472	2801.1	3055.8	6.1516
680	10	37.145	0.026922	2849.2	3118.4	6.2451
700	10	35.355	0.028285	2894.5	3177.4	6.3305
720	10	33.804	0.029582	2937.8	3233.7	6.4098
740	10	32.437	0.030829	2979.7	3288.0	6.4843
760	10	31.215	0.032036	3020.6	3341.0	6.5549
780	10	30.112	0.033209	3060.7	3392.8	6.6222
800	10	29.107	0.034356	3100.2	3443.7	6.6867
820	10	28.185	0.035479	3139.3	3494.1	6.7489
840	10	27.335	0.036584	3178.1	3543.9	6.8089
860	10	26.546	0.037671	3216.7	3593.4	6.8671
880	10	25.81	0.038744	3255.1	3642.6	6.9237
900	10	25.123	0.039804	3293.5	3691.6	6.9787
920	10	24.478	0.040854	3331.9	3740.4	7.0324
940	10	23.87	0.041893	3370.3	3789.2	7.0849
960	10	23.297	0.042924	3408.7	3837.9	7.1362
980	10	22.755	0.043947	3447.2	3886.7	7.1864
1000	10	22.241	0.044963	3485.8	3935.5	7.2357
1020	10	21.752	0.045972	3524.6	3984.3	7.284
1040	10	21.287	0.046976	3563.4	4033.2	7.3315
1060	10	20.844	0.047975	3602.5	4082.2	7.3782
1080	10	20.421	0.048969	3641.6	4131.3	7.4241
1100	10	20.017	0.049959	3681.0	4180.6	7.4693

Table D.3Q   Isobaric Data for $P = 12.0$ MPa

T (K)	P (MPa)	ρ (kg/m³)	v (m³/kg)	u (kJ/kg)	h (kJ/kg)	s (kJ/kg·K)
597.83	12	70.106	0.014264	2514.3	2685.4	5.4939
600	12	68.549	0.014588	2528.8	2703.8	5.5247
620	12	59.113	0.016917	2628.8	2831.8	5.7346
640	12	53.502	0.018691	2701.7	2926.0	5.8843
660	12	49.499	0.020202	2762.6	3005.0	6.0059
680	12	46.392	0.021555	2816.6	3075.3	6.1108
700	12	43.857	0.022801	2866.3	3139.9	6.2045
720	12	41.718	0.023971	2912.9	3200.5	6.2899
740	12	39.87	0.025082	2957.4	3258.4	6.3692
760	12	38.245	0.026147	3000.4	3314.2	6.4436
780	12	36.796	0.027177	3042.3	3368.4	6.514
800	12	35.49	0.028177	3083.3	3421.4	6.5811
820	12	34.303	0.029152	3123.7	3473.5	6.6454
840	12	33.215	0.030107	3163.6	3524.9	6.7073
860	12	32.213	0.031044	3203.2	3575.7	6.7671
880	12	31.284	0.031966	3242.5	3626.1	6.8251
900	12	30.419	0.032874	3281.7	3676.2	6.8813
920	12	29.611	0.033771	3320.7	3726.0	6.9361
940	12	28.853	0.034658	3359.7	3775.6	6.9895
960	12	28.14	0.035536	3398.7	3825.2	7.0416
980	12	27.468	0.036406	3437.8	3874.6	7.0926
1000	12	26.832	0.037269	3476.9	3924.1	7.1425
1020	12	26.229	0.038126	3516.0	3973.5	7.1915
1040	12	25.657	0.038976	3555.3	4023.0	7.2395
1060	12	25.112	0.039821	3594.7	4072.5	7.2867
1080	12	24.593	0.040662	3634.2	4122.1	7.3331
1100	12	24.098	0.041498	3673.9	4171.8	7.3787

Table D.3R   Isobaric Data for $P = 14.0$ MPa

T (K)	P (MPa)	ρ (kg/m³)	v (m³/kg)	u (kJ/kg)	h (kJ/kg)	s (kJ/kg·K)
609.82	14	87.069	0.011485	2477.1	2637.9	5.3727
620	14	77.66	0.012877	2549.9	2730.1	5.5229
640	14	67.43	0.01483	2646.6	2854.2	5.72
660	14	61.128	0.016359	2719.6	2948.6	5.8653
680	14	56.586	0.017672	2781.1	3028.5	5.9847
700	14	53.047	0.018851	2836.0	3099.9	6.0882
720	14	50.153	0.019939	2886.5	3165.7	6.1808
740	14	47.711	0.020959	2934.1	3227.5	6.2655
760	14	45.602	0.021929	2979.5	3286.5	6.3441
780	14	43.747	0.022859	3023.3	3343.3	6.418
800	14	42.094	0.023756	3066.0	3398.5	6.4879
820	14	40.605	0.024628	3107.7	3452.5	6.5545
840	14	39.252	0.025477	3148.8	3505.5	6.6184
860	14	38.012	0.026307	3189.5	3557.8	6.6799
880	14	36.871	0.027122	3229.7	3609.4	6.7392
900	14	35.813	0.027923	3269.7	3660.6	6.7967
920	14	34.829	0.028712	3309.5	3711.4	6.8526
940	14	33.91	0.02949	3349.1	3762.0	6.9069
960	14	33.048	0.030259	3388.7	3812.3	6.96
980	14	32.237	0.03102	3428.2	3862.5	7.0117
1000	14	31.472	0.031774	3467.8	3912.6	7.0623
1020	14	30.749	0.032521	3507.4	3962.7	7.1119
1040	14	30.064	0.033262	3547.1	4012.8	7.1605
1060	14	29.414	0.033998	3586.8	4062.8	7.2082
1080	14	28.795	0.034729	3626.7	4112.9	7.255
1100	14	28.205	0.035455	3666.7	4163.1	7.301
1120	14	27.641	0.036178	3706.8	4213.3	7.3463

Table D.3S   Isobaric Data for $P = 16.0$ MPa

T (K)	P (MPa)	ρ (kg/m³)	v (m³/kg)	u (kJ/kg)	h (kJ/kg)	s (kJ/kg·K)
620.5	16	107.42	0.009309	2431.8	2580.8	5.2463
640	16	85.058	0.011757	2579.1	2767.2	5.5427
660	16	74.683	0.01339	2670.5	2884.8	5.7237
680	16	67.984	0.014709	2742.1	2977.5	5.8622
700	16	63.065	0.015857	2803.5	3057.2	5.9778
720	16	59.195	0.016893	2858.6	3128.9	6.0788
740	16	56.014	0.017853	2909.6	3195.3	6.1697
760	16	53.32	0.018755	2957.7	3257.8	6.253
780	16	50.988	0.019613	3003.7	3317.5	6.3306
800	16	48.935	0.020435	3048.1	3375.1	6.4035
820	16	47.103	0.021230	3091.4	3431.1	6.4726
840	16	45.452	0.022001	3133.8	3485.8	6.5386
860	16	43.951	0.022753	3175.5	3539.5	6.6018
880	16	42.576	0.023488	3216.7	3592.5	6.6627
900	16	41.308	0.024208	3257.5	3644.8	6.7215
920	16	40.134	0.024916	3298.0	3696.7	6.7785
940	16	39.042	0.025614	3338.3	3748.2	6.8338
960	16	38.021	0.026301	3378.5	3799.4	6.8877
980	16	37.063	0.026981	3418.6	3850.3	6.9403
1000	16	36.163	0.027653	3458.7	3901.1	6.9916
1020	16	35.313	0.028318	3498.8	3951.8	7.0418
1040	16	34.51	0.028977	3538.8	4002.5	7.091
1060	16	33.749	0.029631	3579.0	4053.1	7.1391
1080	16	33.026	0.030279	3619.2	4103.7	7.1864
1100	16	32.338	0.030924	3659.5	4154.3	7.2329
1120	16	31.682	0.031564	3700.0	4205.0	7.2785
1140	16	31.056	0.0322	3740.5	4255.7	7.3235

Table D.3T   Isobaric Data for $P = 18.0$ MPa

T (K)	P (MPa)	ρ (kg/m³)	v (m³/kg)	u (kJ/kg)	h (kJ/kg)	s (kJ/kg·K)
630.14	18	133.3	0.0075017	2374.8	2509.8	5.1061
640	18	110	0.0090911	2489.2	2652.8	5.3314
660	18	91.08	0.010979	2613.3	2810.9	5.5749
680	18	80.955	0.012353	2698.9	2921.2	5.7397
700	18	74.095	0.013496	2768.4	3011.3	5.8704
720	18	68.943	0.014505	2829.0	3090.1	5.9813
740	18	64.839	0.015423	2883.9	3161.6	6.0793
760	18	61.439	0.016276	2935.0	3228.0	6.1679
780	18	58.544	0.017081	2983.4	3290.9	6.2495
800	18	56.029	0.017848	3029.8	3351.0	6.3257
820	18	53.809	0.018584	3074.7	3409.2	6.3975
840	18	51.825	0.019296	3118.4	3465.7	6.4656
860	18	50.034	0.019986	3161.3	3521.0	6.5307
880	18	48.403	0.02066	3203.5	3575.3	6.5931
900	18	46.908	0.021318	3245.2	3628.9	6.6533
920	18	45.529	0.021964	3286.5	3681.8	6.7115
940	18	44.251	0.022598	3327.5	3734.3	6.7679
960	18	43.06	0.023223	3368.3	3786.3	6.8227
980	18	41.948	0.023839	3409.0	3838.1	6.876
1000	18	40.904	0.024448	3449.5	3889.6	6.9281
1020	18	39.922	0.025049	3490.0	3940.9	6.9789
1040	18	38.995	0.025645	3530.6	3992.2	7.0286
1060	18	38.118	0.026234	3571.1	4043.3	7.0774
1080	18	37.287	0.026819	3611.7	4094.4	7.1251
1100	18	36.497	0.0274	3652.3	4145.5	7.172
1120	18	35.745	0.027976	3693.1	4196.6	7.2181
1140	18	35.029	0.028548	3733.9	4247.8	7.2633

Table D.3U   Isobaric Data for P = 20.0 MPa

T (K)	P (MPa)	ρ (kg/m³)	v (m³/kg)	u (kJ/kg)	h (kJ/kg)	s (kJ/kg·K)
638.9	20	170.5	0.005865	2295.0	2412.3	4.9314
640	20	160.5	0.006231	2328.4	2453.0	4.995
660	20	112.09	0.008921	2543.7	2722.1	5.4104
680	20	96.059	0.01041	2650.1	2858.3	5.6139
700	20	86.38	0.011577	2730.2	2961.8	5.7639
720	20	79.525	0.012575	2797.4	3048.9	5.8867
740	20	74.257	0.013467	2857.0	3126.3	5.9927
760	20	70.002	0.014285	2911.5	3197.2	6.0872
780	20	66.444	0.01505	2962.5	3263.5	6.1733
800	20	63.396	0.015774	3010.9	3326.4	6.253
820	20	60.736	0.016465	3057.5	3386.8	6.3276
840	20	58.379	0.017129	3102.7	3445.3	6.398
860	20	56.268	0.017772	3146.8	3502.2	6.465
880	20	54.357	0.018397	3190.1	3558.0	6.5291
900	20	52.615	0.019006	3232.7	3612.8	6.5907
920	20	51.015	0.019602	3274.8	3666.8	6.6501
940	20	49.538	0.020186	3316.5	3720.2	6.7075
960	20	48.168	0.020761	3358.0	3773.2	6.7633
980	20	46.891	0.021326	3399.2	3825.7	6.8174
1000	20	45.696	0.021884	3440.3	3878.0	6.8702
1020	20	44.574	0.022435	3481.3	3930.0	6.9217
1040	20	43.518	0.022979	3522.2	3981.8	6.972
1060	20	42.521	0.023518	3563.2	4033.5	7.0213
1080	20	41.577	0.024052	3604.1	4085.1	7.0695
1100	20	40.682	0.024581	3645.1	4136.7	7.1168
1120	20	39.831	0.025106	3686.1	4188.3	7.1633
1140	20	39.021	0.025627	3727.3	4239.8	7.2089

Table D.3V   Isobaric Data for P = 26.0 MPa (Supercritical)

T (K)	P (MPa)	ρ (kg/m³)	v (m³/kg)	u (kJ/kg)	h (kJ/kg)	s (kJ/kg·K)
660	26	365.88	0.002733	2010.9	2082.0	4.386
680	26	167.08	0.005985	2448.8	2604.4	5.1692
700	26	134.79	0.007419	2591.2	2784.1	5.43
720	26	118.02	0.008473	2688.8	2909.1	5.6062
740	26	106.98	0.009348	2767.2	3010.2	5.7447
760	26	98.863	0.010115	2834.6	3097.6	5.8613
780	26	92.512	0.010809	2895.3	3176.3	5.9635
800	26	87.325	0.011451	2951.2	3248.9	6.0554
820	26	82.963	0.012054	3003.7	3317.1	6.1396
840	26	79.211	0.012625	3053.7	3382.0	6.2178
860	26	75.927	0.01317	3101.9	3444.3	6.2912
880	26	73.015	0.013696	3148.6	3504.7	6.3606
900	26	70.403	0.014204	3194.2	3563.5	6.4267
920	26	68.039	0.014697	3239.0	3621.1	6.4899
940	26	65.882	0.015179	3283.0	3677.6	6.5507
960	26	63.902	0.015649	3326.5	3733.3	6.6094
980	26	62.074	0.01611	3369.5	3788.4	6.6661
1000	26	60.378	0.016562	3412.2	3842.9	6.7211
1020	26	58.797	0.017008	3454.7	3896.9	6.7747
1040	26	57.318	0.017447	3497.0	3950.6	6.8268
1060	26	55.93	0.01788	3539.2	4004.0	6.8777
1080	26	54.623	0.018307	3581.2	4057.2	6.9274
1100	26	53.389	0.01873	3623.3	4110.2	6.976
1120	26	52.222	0.019149	3665.3	4163.1	7.0237
1140	26	51.115	0.019564	3707.3	4216.0	7.0704
1160	26	50.062	0.019975	3749.4	4268.7	7.1163
1180	26	49.06	0.020383	3791.5	4321.4	7.1614
1200	26	48.104	0.020788	3833.7	4374.2	7.2057

Table D.3W   Isobaric Data for $P = 28.0$ MPa (Supercritical)

T (K)	P (MPa)	ρ (kg/m³)	v (m³/kg)	u (kJ/kg)	h (kJ/kg)	s (kJ/kg·K)
660	28	455.47	0.0021955	1897.4	1958.9	4.1923
680	28	210.73	0.0047453	2347.1	2480.0	4.9705
700	28	157.04	0.0063676	2533.7	2712.0	5.3072
720	28	133.92	0.0074672	2647.0	2856.1	5.5104
740	28	119.74	0.0083511	2733.9	2967.7	5.6634
760	28	109.74	0.0091127	2806.9	3062.0	5.7892
780	28	102.1	0.0097942	2871.3	3145.6	5.8977
800	28	95.978	0.010419	2930.1	3221.9	5.9943
820	28	90.896	0.011002	2984.9	3293.0	6.0821
840	28	86.571	0.011551	3036.8	3360.2	6.1632
860	28	82.818	0.012075	3086.5	3424.6	6.2389
880	28	79.511	0.012577	3134.5	3486.6	6.3102
900	28	76.563	0.013061	3181.1	3546.8	6.3779
920	28	73.906	0.013531	3226.8	3605.6	6.4425
940	28	71.494	0.013987	3271.6	3663.2	6.5044
960	28	69.286	0.014433	3315.8	3719.9	6.5641
980	28	67.254	0.014869	3359.5	3775.8	6.6217
1000	28	65.374	0.015297	3402.8	3831.1	6.6775
1020	28	63.626	0.015717	3445.8	3885.8	6.7318
1040	28	61.994	0.016131	3488.5	3940.2	6.7845
1060	28	60.465	0.016538	3531.1	3994.2	6.8359
1080	28	59.029	0.016941	3573.6	4047.9	6.8862
1100	28	57.675	0.017339	3615.9	4101.4	6.9353
1120	28	56.395	0.017732	3658.3	4154.8	6.9833
1140	28	55.183	0.018121	3700.6	4208.0	7.0304
1160	28	54.033	0.018507	3743.0	4261.2	7.0767
1180	28	52.938	0.01889	3785.3	4314.3	7.122
1200	28	51.895	0.01927	3827.8	4367.3	7.1666

Table D.3X   Isobaric Data for $P = 32.0$ MPa (Supercritical)

T (K)	P (MPa)	ρ (kg/m³)	v (m³/kg)	u (kJ/kg)	h (kJ/kg)	s (kJ/kg·K)
660	32	516.64	0.001936	1823.6	1885.6	4.0689
680	32	351.66	0.002844	2100.2	2191.2	4.5242
700	32	217.87	0.00459	2395.5	2542.3	5.0338
720	32	172.38	0.005801	2553.3	2739.0	5.3111
740	32	148.9	0.006716	2661.8	2876.7	5.4999
760	32	133.76	0.007476	2747.9	2987.1	5.6471
780	32	122.84	0.00814	2821.2	3081.7	5.77
800	32	114.42	0.00874	2886.5	3166.2	5.877
820	32	107.62	0.009292	2946.2	3243.6	5.9726
840	32	101.96	0.009808	3002.0	3315.9	6.0597
860	32	97.13	0.010295	3054.9	3384.4	6.1403
880	32	92.934	0.01076	3105.6	3449.9	6.2156
900	32	89.235	0.011206	3154.5	3513.1	6.2866
920	32	85.935	0.011637	3202.1	3574.4	6.354
940	32	82.962	0.012054	3248.6	3634.3	6.4184
960	32	80.262	0.012459	3294.3	3692.9	6.4802
980	32	77.791	0.012855	3339.3	3750.6	6.5396
1000	32	75.517	0.013242	3383.7	3807.5	6.597
1020	32	73.413	0.013622	3427.7	3863.6	6.6527
1040	32	71.458	0.013994	3471.5	3919.3	6.7067
1060	32	69.633	0.014361	3514.9	3974.4	6.7592
1080	32	67.924	0.014722	3558.1	4029.3	6.8105
1100	32	66.318	0.015079	3601.2	4083.8	6.8605
1120	32	64.804	0.015431	3644.3	4138.0	6.9094
1140	32	63.374	0.015779	3687.2	4192.1	6.9572
1160	32	62.019	0.016124	3730.1	4246.1	7.0041
1180	32	60.734	0.016465	3773.0	4299.9	7.0502
1200	32	59.511	0.016804	3816.0	4353.7	7.0953

Table D.4 Compressed Liquid (Water)*

Table D.4A Isobaric Data for P = 5.0 MPa

T (K)	P (MPa)	ρ (kg/m³)	v (m³/kg)	u (kJ/kg)	h (kJ/kg)	s (kJ/kg·K)
273.15	5	1002.3	0.000998	0.044068	5.0	0.0001
280	5	1002.3	0.000998	28.731	33.7	0.1039
300	5	998.74	0.001001	112.15	117.2	0.3917
320	5	991.56	0.001009	195.5	200.5	0.66066
340	5	981.68	0.001019	278.9	284.0	0.91364
360	5	969.62	0.001031	362.5	367.7	1.1528
380	5	955.65	0.001046	446.5	451.7	1.38
400	5	939.91	0.001064	530.9	536.2	1.5968
420	5	922.46	0.001084	616.1	621.5	1.8047
440	5	903.27	0.001107	702.2	707.7	2.0053
460	5	882.22	0.001134	789.6	795.2	2.1998
480	5	859.08	0.001164	878.6	884.5	2.3897
500	5	833.51	0.001200	970.0	976.0	2.5764
520	5	804.92	0.001242	1064.3	1070.5	2.7618
537.09	5	777.37	0.001286	1148.2	1154.6	2.921

*Property values generated from NIST Database 23: REFPROP Version 7.0 (August 2002).

Table D.4B Isobaric Data for P = 10.0 MPa

T (K)	P (MPa)	ρ (kg/m³)	v (m³/kg)	u (kJ/kg)	h (kJ/kg)	s (kJ/kg·K)
273.15	10	1004.8	0.0009952	0.1171	10.069	0.0003376
280	10	1004.7	0.0009954	28.659	38.613	0.10355
300	10	1001.0	0.0009991	111.74	121.73	0.39029
320	10	993.7	0.0010063	194.78	204.84	0.65846
340	10	983.84	0.0010164	277.92	288.08	0.91079
360	10	971.85	0.001029	361.27	371.56	1.1493
380	10	957.99	0.0010439	444.93	455.37	1.3759
400	10	942.42	0.0010611	529.06	539.67	1.5921
420	10	925.19	0.0010809	613.84	624.65	1.7994
440	10	906.28	0.0011034	699.52	710.55	1.9992
460	10	885.59	0.0011292	786.38	797.67	2.1928
480	10	862.94	0.0011588	874.8	886.39	2.3816
500	10	838.02	0.0011933	965.25	977.18	2.5669
520	10	810.36	0.001234	1058.4	1070.7	2.7504
540	10	779.15	0.0012835	1155.3	1168.1	2.9341
560	10	743.0	0.0013459	1257.6	1271.1	3.1213
580	10	699.05	0.0014305	1368.8	1383.1	3.3177
584.15	10	688.42	0.0014526	1393.5	1408.1	3.3606

Table D.4C   Isobaric Data for P = 15.0 MPa

T (K)	P (MPa)	ρ (kg/m³)	ν (m³/kg)	u (kJ/kg)	h (kJ/kg)	s (kJ/kg·K)
273.15	15	1007.3	0.000993	0.17746	15.069	0.00044686
280	15	1007.0	0.000993	28.581	43.476	0.10316
300	15	1003.1	0.000997	111.34	126.29	0.38886
320	15	995.83	0.001004	194.1	209.16	0.65627
340	15	985.98	0.001014	276.99	292.2	0.90797
360	15	974.05	0.001027	360.07	375.47	1.1459
380	15	960.3	0.001041	443.45	459.07	1.3719
400	15	944.88	0.001058	527.26	543.13	1.5875
420	15	927.86	0.001078	611.68	627.85	1.7942
440	15	909.22	0.0011	696.94	713.44	1.9933
460	15	888.87	0.001125	783.3	800.18	2.186
480	15	866.68	0.001154	871.1	888.4	2.3738
500	15	842.36	0.001187	960.75	978.56	2.5578
520	15	815.52	0.001226	1052.8	1071.2	2.7395
540	15	785.52	0.001273	1148.2	1167.3	2.9208
560	15	751.28	0.001331	1248.3	1268.2	3.1042
580	15	710.78	0.001407	1355.3	1376.4	3.294
600	15	659.41	0.001517	1474.9	1497.7	3.4994
615.31	15	603.52	0.001657	1585.3	1610.2	3.6846

Table D.4D   Isobaric Data for P = 20.0 MPa

T (K)	P (MPa)	ρ (kg/m³)	ν (m³/kg)	u (kJ/kg)	h (kJ/kg)	s (kJ/kg·K)
273.15	20	1009.7	0.00099	0.22569	20.033	0.00046962
280	20	1009.4	0.000991	28.496	48.31	0.10271
300	20	1005.3	0.000995	110.94	130.84	0.38741
320	20	997.93	0.001002	193.44	213.48	0.65408
340	20	988.09	0.001012	276.07	296.31	0.90516
360	20	976.22	0.001024	358.89	379.38	1.1426
380	20	962.58	0.001039	442.0	462.77	1.368
400	20	947.31	0.001056	525.5	546.62	1.583
420	20	930.48	0.001075	609.58	631.08	1.7891
440	20	912.09	0.001096	694.44	716.37	1.9874
460	20	892.08	0.001121	780.32	802.74	2.1794
480	20	870.3	0.001149	867.53	890.51	2.3662
500	20	846.53	0.001181	956.44	980.07	2.5489
520	20	820.44	0.001219	1047.6	1071.9	2.7291
540	20	791.5	0.001263	1141.6	1166.9	2.9082
560	20	758.86	0.001318	1239.7	1266.0	3.0885
580	20	721.04	0.001387	1343.5	1371.3	3.2731
600	20	675.11	0.001481	1456.8	1486.4	3.4682
620	20	613.23	0.001631	1588.6	1621.2	3.6891
638.9	20	490.19	0.00204	1786.4	1827.2	4.0156

Table D.4E   Isobaric Data for P = 30.0 MPa

T (K)	P (MPa)	ρ (kg/m^3)	v (m^3/kg)	u (kJ/kg)	h (kJ/kg)	s (kJ/kg·K)
273.15	30	1014.5	0.0009857	0.28791	29.858	0.0002688
280	30	1014.0	0.0009862	28.307	57.893	0.10164
300	30	1009.6	0.0009905	110.16	139.87	0.38444
320	30	1002.1	0.0009979	192.15	222.08	0.64972
340	30	992.24	0.0010078	274.29	304.52	0.89961
360	30	980.48	0.0010199	356.62	387.21	1.1359
380	30	967.04	0.0010341	439.19	470.21	1.3603
400	30	952.05	0.0010504	522.11	553.62	1.5742
420	30	935.59	0.0010688	605.53	637.6	1.7791
440	30	917.67	0.0010897	689.63	722.33	1.9761
460	30	898.25	0.0011133	774.62	808.02	2.1666
480	30	877.23	0.00114	860.75	894.95	2.3516
500	30	854.45	0.0011703	948.32	983.43	2.5322
520	30	829.66	0.0012053	1037.7	1073.9	2.7095
540	30	802.5	0.0012461	1129.5	1166.9	2.885
560	30	772.42	0.0012946	1224.4	1263.2	3.0601
580	30	738.56	0.001354	1323.4	1364.0	3.237
600	30	699.47	0.0014296	1428.5	1471.4	3.4189
620	30	652.41	0.0015328	1542.9	1588.9	3.6116
640	30	591.01	0.001692	1674.8	1725.5	3.8283

Table D.4F   Isobaric Data for P = 50.0 MPa

T (K)	P (MPa)	ρ (kg/m^3)	v (m^3/kg)	u (kJ/kg)	h (kJ/kg)	s (kJ/kg·K)
273.15	50	1023.8	0.000977	0.28922	49.126	−0.0010315
280	50	1022.9	0.000978	27.862	76.742	0.098824
300	50	1017.8	0.000982	108.63	157.76	0.37828
320	50	1010.1	0.00099	189.68	239.18	0.64102
340	50	1000.3	0.001	270.92	320.9	0.88875
360	50	988.72	0.001011	352.32	402.89	1.123
380	50	975.62	0.001025	433.9	485.15	1.3454
400	50	961.12	0.00104	515.76	567.78	1.5573
420	50	945.31	0.001058	597.99	650.88	1.7601
440	50	928.2	0.001077	680.73	734.6	1.9548
460	50	909.8	0.001099	764.15	819.1	2.1426
480	50	890.05	0.001124	848.42	904.59	2.3245
500	50	868.88	0.001151	933.75	991.29	2.5015
520	50	846.15	0.001182	1020.4	1079.5	2.6744
540	50	821.66	0.001217	1108.6	1169.5	2.8442
560	50	795.16	0.001258	1198.9	1261.8	3.012
580	50	766.3	0.001305	1291.7	1356.9	3.1789
600	50	734.55	0.001361	1387.6	1455.7	3.3463
620	50	699.21	0.00143	1487.7	1559.2	3.516
640	50	659.19	0.001517	1593.4	1669.3	3.6907

Table D.5 Vapor Properties: Saturated Solid (Ice)–Water Vapor

Temp., $T\,°C$	Sat. press., P_{sat} kPa	Sat. ice, $v_i \times 10^3$	Sat. vapor, v_g	Sat. ice, u_i	Subl., u_{ig}	Sat. vapor, u_g	Sat. ice, h_i	Subl., h_{ig}	Sat. vapor, h_g	Sat. ice, s_i	Subl., s_{ig}	Sat. vapor, s_g
		Specific volume, m³/kg		Internal energy, kJ/kg			Enthalpy, kJ/kg			Entropy, kJ/kg·K		
0.01	0.6113	1.0908	206.1	−333.40	2708.7	2375.3	−333.40	2834.8	2501.4	−1.221	10.378	9.156
0	0.6108	1.0908	206.3	−333.43	2708.8	2375.3	−333.43	2834.8	2501.3	−1.221	10.378	9.157
−2	0.5178	1.0904	241.7	−337.62	2710.2	2372.6	−337.62	2835.3	2497.7	−1.237	10.456	9.219
−4	0.4375	1.0901	283.8	−341.78	2711.6	2369.8	−341.78	2835.7	2494.0	−1.253	10.536	9.283
−6	0.3689	1.0898	334.2	−345.91	2712.9	2367.0	−345.91	2836.3	2490.3	−1.268	10.616	9.348
−8	0.3102	1.0894	394.4	−350.02	2714.2	2364.2	−350.02	2836.6	2486.6	−1.284	10.698	9.414
−10	0.2602	1.0891	466.7	−354.09	2715.5	2361.4	−354.09	2837.0	2482.9	−1.299	10.781	9.481
−12	0.2176	1.0888	553.7	−358.14	2716.8	2358.7	−358.14	2837.3	2479.2	−1.315	10.865	9.550
−14	0.1815	1.0884	658.8	−362.15	2718.0	2355.9	−362.15	2837.6	2475.5	−1.331	10.950	9.619
−16	0.1510	1.0881	786.0	−366.14	2719.2	2353.1	−366.14	2837.9	2471.8	−1.346	11.036	9.690
−18	0.1252	1.0878	940.5	−370.10	2720.4	2350.3	−370.10	2838.2	2468.1	−1.362	11.123	9.762
−20	0.1035	1.0874	1128.6	−374.03	2721.6	2347.5	−374.03	2838.4	2464.3	−1.377	11.212	9.835
−22	0.0853	1.0871	1358.4	−377.93	2722.7	2344.7	−377.93	2838.6	2460.6	−1.393	11.302	9.909
−24	0.0701	1.0868	1640.1	−381.80	2723.7	2342.0	−381.80	2838.7	2456.9	−1.408	11.394	9.985
−26	0.0574	1.0864	1986.4	−385.64	2724.8	2339.2	−385.64	2838.9	2453.2	−1.424	11.486	10.062
−28	0.0469	1.0861	2413.7	−389.45	2725.8	2336.4	−389.45	2839.0	2449.5	−1.439	11.580	10.141
−30	0.0381	1.0858	2943	−393.23	2726.8	2333.6	−393.23	2839.0	2445.8	−1.455	11.676	10.221
−32	0.0309	1.0854	3600	−396.98	2727.8	2330.8	−396.98	2839.1	2442.1	−1.471	11.773	10.303
−34	0.0250	1.0851	4419	−400.71	2728.7	2328.0	−400.71	2839.1	2438.4	−1.486	11.872	10.386
−36	0.0201	1.0848	5444	−404.40	2729.6	2325.2	−404.40	2839.1	2434.7	−1.501	11.972	10.470
−38	0.0161	1.0844	6731	−408.06	2730.5	2322.4	−408.06	2839.9	2430.9	−1.517	12.073	10.556
−40	0.0129	1.0841	8354	−411.70	2731.3	2319.6	−411.70	2839.9	2427.2	−1.532	12.176	10.644

Appendix E
Various Thermodynamic Data

Table E.1 Critical Constants and Specific Heats for Selected Gases*

Substance	$\mathcal{M}$ (kg/kmol)	T_c (K)	P_c (10^5 Pa)	$\bar{v}_c$ (m³/kmol)	Z_c	c_v (kJ/kg·K)	c_p (kJ/kg·K)
Acetylene (C_2H_2)	26.04	309	62.4	0.112	0.272	1.37	1.69
Air (equivalent)	28.97	133	37.7	0.0829	0.284	0.718	1.005
Ammonia (NH_3)	17.04	406	112.8	0.0723	0.242	1.66	2.15
Benzene (C_6H_6)	78.11	562	48.3	0.256	0.274	0.67	0.775
n-Butane (C_4H_{10})	58.12	425.2	37.9	0.257	0.274	1.56	1.71
Carbon dioxide (CO_2)	44.01	304.2	73.9	0.0941	0.276	0.657	0.846
Carbon monoxide (CO)	28.01	133	35.0	0.0928	0.294	0.744	1.04
Refrigerant 134a ($C_2F_4H_2$)	102.03	374.3	40.6	0.200	0.262	0.76	0.85
Ethane (C_2H_6)	30.07	305.4	48.8	0.148	0.285	1.48	1.75
Ethylene (C_2H_4)	28.05	283	51.2	0.128	0.279	1.23	1.53
Helium (He)	4.003	5.2	2.3	0.0579	0.300	3.12	5.19
Hydrogen (H_2)	2.016	33.2	13.0	0.0648	0.304	10.2	14.3
Methane (CH_4)	16.04	190.7	46.4	0.0991	0.290	1.70	2.22
Nitrogen (N_2)	28.01	126.2	33.9	0.0897	0.291	0.743	1.04
Oxygen (O_2)	32.00	154.4	50.5	0.0741	0.290	0.658	0.918
Propane (C_3H_8)	44.09	370	42.5	0.200	0.278	1.48	1.67
Sulfur dioxide (SO_2)	64.06	431	78.7	0.124	0.268	0.471	0.601
Water (H_2O)	18.02	647.1	220.6	0.0558	0.230	1.40	1.86

*Adapted from Wark, K., Jr., and Richards, D. E., *Thermodynamics,* 6th ed., McGraw-Hill, New York, 1999.

Table E.2 Van der Waals Constants for Selected Gases*

Substance	a [10^5 Pa·(m³/kmol)²]	b (m³/kmol)	Substance	a [10^5 Pa·(m³/kmol)²]	b (m³/kmol)
Acetylene (C_2H_2)	4.410	0.0510	Ethylene (C_2H_4)	4.563	0.0574
Air (equivalent)	1.358	0.0364	Helium (He)	0.0341	0.0234
Ammonia (NH_3)	4.223	0.0373	Hydrogen (H_2)	0.247	0.0265
Benzene (C_6H_6)	18.63	0.1181	Methane (CH_4)	2.285	0.0427
n-Butane (C_4H_{10})	13.80	0.1196	Nitrogen (N_2)	1.361	0.0385
Carbon dioxide (CO_2)	3.643	0.0427	Oxygen (O_2)	1.369	0.0315
Carbon monoxide (CO)	1.463	0.0394	Propane (C_3H_8)	9.315	0.0900
Refrigerant 134a ($C_2F_4H_2$)	10.05	0.0957	Sulfur dioxide (SO_2)	6.837	0.0568
Ethane (C_2H_6)	5.575	0.0650	Water (H_2O)	5.507	0.0304

*Adapted from Wark, K., Jr., and Richards, D. E., *Thermodynamics,* 6th ed., McGraw-Hill, New York, 1999.

Appendix F
Thermo-Physical Properties of Selected Gases at 1 atm

Table F.1   Thermo-Physical Properties of Selected Gases (1 atm)

Table F.1A   Ammonia (NH_3)*

T (K)	ρ (kg/m³)	c_p (kJ/kg·K)	μ (µPa·s)	ν (m²/s)	k (W/m·K)	α (m²/s)	Pr
239.824	0.8895	2.297	8.054	9.054E-06	0.0210	1.026E-05	0.8822
240	0.8888	2.296	8.059	9.068E-06	0.0210	1.028E-05	0.8820
260	0.8135	2.207	8.734	1.074E-05	0.0220	1.228E-05	0.8744
280	0.7515	2.172	9.436	1.256E-05	0.0234	1.435E-05	0.8748
300	0.6990	2.165	10.160	1.454E-05	0.0251	1.659E-05	0.8762
320	0.6538	2.174	10.902	1.668E-05	0.0271	1.904E-05	0.8759
340	0.6143	2.193	11.657	1.898E-05	0.0293	2.173E-05	0.8734
360	0.5795	2.219	12.422	2.144E-05	0.0317	2.467E-05	0.8691
380	0.5485	2.249	13.195	2.406E-05	0.0344	2.786E-05	0.8634
400	0.5207	2.283	13.971	2.683E-05	0.0372	3.130E-05	0.8572
420	0.4956	2.320	14.751	2.976E-05	0.0402	3.498E-05	0.8510
440	0.4728	2.358	15.531	3.285E-05	0.0433	3.886E-05	0.8453
460	0.4521	2.399	16.310	3.607E-05	0.0465	4.292E-05	0.8405
480	0.4331	2.440	17.088	3.945E-05	0.0498	4.714E-05	0.8369
500	0.4157	2.483	17.863	4.297E-05	0.0531	5.147E-05	0.8349
520	0.3996	2.526	18.635	4.663E-05	0.0564	5.587E-05	0.8346
540	0.3848	2.570	19.403	5.043E-05	0.0596	6.030E-05	0.8362
560	0.3710	2.615	20.167	5.436E-05	0.0628	6.471E-05	0.8401
580	0.3581	2.660	20.927	5.843E-05	0.0658	6.904E-05	0.8463
600	0.3462	2.706	21.682	6.264E-05	0.0686	7.324E-05	0.8552

*Property values generated from NIST Database 23: REFPROP Version 7.0 (August 2002).

Table F.1B   Carbon Dioxide (CO_2)*

T (K)	ρ (kg/m³)	c_p (kJ/kg·K)	μ (µPa·s)	v (m²/s)	k (W/m·K)	α (m²/s)	Pr
220	2.472	0.781	11.06	4.475E-06	0.0109	5.647E-06	0.792
240	2.258	0.796	12.07	5.344E-06	0.0122	6.808E-06	0.785
260	2.079	0.814	13.06	6.282E-06	0.0137	8.079E-06	0.778
280	1.927	0.833	14.05	7.288E-06	0.0152	9.465E-06	0.770
300	1.797	0.853	15.02	8.361E-06	0.0168	1.096E-05	0.763
320	1.683	0.872	15.98	9.499E-06	0.0184	1.256E-05	0.756
340	1.583	0.890	16.93	1.070E-05	0.0201	1.426E-05	0.750
360	1.494	0.908	17.87	1.196E-05	0.0218	1.605E-05	0.745
380	1.415	0.925	18.79	1.328E-05	0.0235	1.792E-05	0.741
400	1.343	0.942	19.70	1.466E-05	0.0251	1.988E-05	0.738
420	1.279	0.958	20.59	1.610E-05	0.0268	2.190E-05	0.735
440	1.220	0.973	21.47	1.759E-05	0.0285	2.401E-05	0.733
460	1.167	0.988	22.33	1.913E-05	0.0302	2.618E-05	0.731
480	1.118	1.002	23.18	2.073E-05	0.0318	2.842E-05	0.729
500	1.073	1.015	24.02	2.237E-05	0.0335	3.072E-05	0.728
520	1.032	1.029	24.84	2.407E-05	0.0351	3.309E-05	0.727
540	0.994	1.041	25.65	2.581E-05	0.0368	3.553E-05	0.727
560	0.958	1.053	26.44	2.760E-05	0.0384	3.802E-05	0.726
580	0.925	1.065	27.23	2.944E-05	0.0400	4.057E-05	0.726
600	0.894	1.076	28.00	3.131E-05	0.0416	4.318E-05	0.725
620	0.865	1.087	28.76	3.324E-05	0.0431	4.585E-05	0.725
640	0.838	1.098	29.50	3.520E-05	0.0447	4.858E-05	0.725
660	0.813	1.108	30.24	3.721E-05	0.0462	5.136E-05	0.724
680	0.789	1.117	30.96	3.925E-05	0.0478	5.420E-05	0.724
700	0.766	1.127	31.68	4.134E-05	0.0493	5.709E-05	0.724
720	0.745	1.136	32.38	4.347E-05	0.0508	6.004E-05	0.724
740	0.725	1.145	33.07	4.563E-05	0.0523	6.304E-05	0.724
760	0.706	1.153	33.75	4.783E-05	0.0538	6.609E-05	0.724
780	0.688	1.161	34.43	5.007E-05	0.0553	6.920E-05	0.724
800	0.670	1.169	35.09	5.234E-05	0.0567	7.236E-05	0.723

*Property values generated from NIST Database 23: REFPROP Version 7.0 (August 2002).

Table F.1C   Carbon Monoxide (CO)*

T (K)	ρ (kg/m³)	c_p (kJ/kg·K)	μ (μPa·s)	ν (m²/s)	k (W/m·K)	α (m²/s)	Pr
200	1.7112	1.0443	12.8977	7.537E-06	0.01923	1.076E-05	0.701
220	1.5544	1.0430	13.9377	8.967E-06	0.02080	1.283E-05	0.699
240	1.4241	1.0420	14.9369	1.049E-05	0.02232	1.504E-05	0.697
260	1.3140	1.0412	15.8998	1.210E-05	0.02378	1.738E-05	0.696
280	1.2198	1.0407	16.8304	1.380E-05	0.02520	1.985E-05	0.695
300	1.1382	1.0402	17.7315	1.558E-05	0.02656	2.243E-05	0.694
320	1.0669	1.0399	18.6057	1.744E-05	0.02789	2.514E-05	0.694
340	1.0040	1.0396	19.4549	1.938E-05	0.02918	2.795E-05	0.693
360	0.9481	1.0394	20.2811	2.139E-05	0.03043	3.088E-05	0.693
380	0.8982	1.0393	21.0859	2.348E-05	0.03165	3.391E-05	0.692
400	0.8532	1.0392	21.8707	2.563E-05	0.03285	3.705E-05	0.692
450	0.7583	1.0392	23.7543	3.133E-05	0.03571	4.532E-05	0.691
500	0.6824	1.0395	25.5404	3.743E-05	0.03844	5.419E-05	0.691
550	0.6204	1.0400	27.2444	4.392E-05	0.04105	6.363E-05	0.690
600	0.5687	1.0408	28.8791	5.078E-05	0.04357	7.362E-05	0.690
650	0.5249	1.0418	30.4548	5.802E-05	0.04601	8.414E-05	0.690
700	0.4874	1.0429	31.9798	6.561E-05	0.04838	9.518E-05	0.689
750	0.4549	1.0443	33.4606	7.355E-05	0.05070	1.067E-04	0.689
800	0.4265	1.0457	34.9026	8.184E-05	0.05297	1.188E-04	0.689

*Property values generated from NIST Database 12: NIST Pure Fluids Version 5.0 (September 2002).

Table F.1D   Helium (He)*

T (K)	ρ (kg/m³)	c_p (kJ/kg·K)	μ (μPa·s)	ν (m²/s)	k (W/m·K)	α (m²/s)	Pr
100	0.487	5.149	9.78	2.008E-05	0.0737	2.914E-05	0.689
120	0.4060	5.1938	10.79	2.659E-05	0.0833	3.952E-05	0.673
140	0.3481	5.1935	11.94	3.432E-05	0.0925	5.117E-05	0.671
160	0.3046	5.1933	13.05	4.284E-05	0.1013	6.404E-05	0.669
180	0.2708	5.1932	14.11	5.212E-05	0.1098	7.806E-05	0.668
200	0.2437	5.1931	15.14	6.213E-05	0.1180	9.322E-05	0.667
220	0.2216	5.1931	16.14	7.286E-05	0.1260	1.095E-04	0.666
240	0.2031	5.1930	17.12	8.429E-05	0.1337	1.268E-04	0.665
260	0.1875	5.1930	18.08	9.641E-05	0.1413	1.451E-04	0.664
280	0.1741	5.1930	19.01	1.092E-04	0.1487	1.645E-04	0.664
300	0.1625	5.1930	19.93	1.226E-04	0.1560	1.848E-04	0.664
320	0.1524	5.1930	20.83	1.367E-04	0.1631	2.061E-04	0.663
340	0.1434	5.1930	21.72	1.514E-04	0.1701	2.284E-04	0.663
360	0.1354	5.1930	22.59	1.668E-04	0.1770	2.516E-04	0.663
380	0.1283	5.1930	23.45	1.827E-04	0.1837	2.757E-04	0.663
400	0.1219	5.1930	24.29	1.993E-04	0.1904	3.007E-04	0.663
420	0.1161	5.1930	25.13	2.164E-04	0.1969	3.266E-04	0.663
440	0.1108	5.1930	25.95	2.342E-04	0.2034	3.534E-04	0.663
460	0.1060	5.1930	26.76	2.525E-04	0.2098	3.811E-04	0.663
480	0.1016	5.1930	27.57	2.714E-04	0.2161	4.096E-04	0.663
500	0.0975	5.1930	28.36	2.908E-04	0.2223	4.389E-04	0.663
550	0.0887	5.1930	30.31	3.419E-04	0.2376	5.159E-04	0.663
600	0.0813	5.1930	32.22	3.963E-04	0.2524	5.980E-04	0.663
650	0.0750	5.1930	34.07	4.541E-04	0.2669	6.850E-04	0.663
700	0.0697	5.1930	35.89	5.152E-04	0.2811	7.768E-04	0.663
750	0.0650	5.1930	37.68	5.794E-04	0.2949	8.733E-04	0.663
800	0.0610	5.1930	39.43	6.468E-04	0.3085	9.745E-04	0.664
850	0.0574	5.1930	41.15	7.172E-04	0.3219	1.080E-03	0.664
900	0.0542	5.1930	42.85	7.907E-04	0.3350	1.190E-03	0.664
950	0.0513	5.1930	44.52	8.671E-04	0.3479	1.305E-03	0.664
1000	0.0488	5.1930	46.16	9.464E-04	0.3606	1.424E-03	0.665

*Property values generated from NIST Database 12: NIST Pure Fluids Version 5.0 (September 2002).

Table F.1E   **Hydrogen (H$_2$)***

T (K)	ρ (kg/m^3)	c_p (kJ/kg·K)	μ (µPa·s)	ν (m^2/s)	k (W/m·K)	α (m^2/s)	Pr
100	0.2457	11.23	4.190	1.705E-05	0.0683	2.477E-05	0.688
150	0.1637	12.61	5.561	3.397E-05	0.1010	4.894E-05	0.694
200	0.1228	13.54	6.780	5.523E-05	0.1324	7.970E-05	0.693
250	0.09820	14.05	7.903	8.047E-05	0.1606	1.164E-04	0.691
300	0.08184	14.31	8.953	1.094E-04	0.1858	1.586E-04	0.690
350	0.07016	14.43	9.946	1.418E-04	0.2103	2.077E-04	0.682
400	0.06139	14.47	10.89	1.774E-04	0.2341	2.634E-04	0.674
450	0.05457	14.49	11.80	2.162E-04	0.2570	3.249E-04	0.665
500	0.04912	14.51	12.67	2.580E-04	0.2805	3.936E-04	0.656
550	0.04465	14.53	13.52	3.027E-04	0.3042	4.689E-04	0.646
600	0.04093	14.54	14.34	3.503E-04	0.3281	5.514E-04	0.635

*Property values generated from NIST Database 12: NIST Pure Fluids Version 5.0 (September 2000).

Table F.1F   **Nitrogen (N$_2$)***

T (K)	ρ (kg/m^3)	c_p (kJ/kg·K)	μ (µPa·s)	ν (m^2/s)	k (W/m·K)	α (m^2/s)	Pr
100	3.4831	1.0718	6.97	2.000E-06	0.0099	2.644E-06	0.756
150	2.2893	1.0486	10.10	4.410E-06	0.0145	6.061E-06	0.728
200	1.7107	1.0435	12.92	7.555E-06	0.0187	1.045E-05	0.723
250	1.3666	1.0418	15.51	1.135E-05	0.0224	1.573E-05	0.721
300	1.1382	1.0414	17.90	1.572E-05	0.0259	2.182E-05	0.721
350	0.9753	1.0423	20.12	2.063E-05	0.0291	2.863E-05	0.721
400	0.8532	1.0450	22.22	2.604E-05	0.0322	3.612E-05	0.721
450	0.7584	1.0497	24.20	3.191E-05	0.0352	4.423E-05	0.721
500	0.6825	1.0564	26.08	3.821E-05	0.0381	5.290E-05	0.722
550	0.6204	1.0650	27.88	4.493E-05	0.0410	6.211E-05	0.723
600	0.5687	1.0751	29.60	5.205E-05	0.0439	7.182E-05	0.725
700	0.4875	1.0981	32.87	6.742E-05	0.0496	9.267E-05	0.728
800	0.4266	1.1223	35.93	8.424E-05	0.0552	1.153E-04	0.731
900	0.3792	1.1457	38.83	1.024E-04	0.0607	1.396E-04	0.733
1000	0.3413	1.1674	41.60	1.219E-04	0.0660	1.656E-04	0.736
1100	0.3103	1.1868	44.25	1.426E-04	0.0712	1.933E-04	0.738
1200	0.2844	1.2040	46.81	1.646E-04	0.0762	2.224E-04	0.740
1300	0.2625	1.2191	49.29	1.878E-04	0.0810	2.532E-04	0.742
1400	0.2438	1.2324	51.70	2.121E-04	0.0858	2.855E-04	0.743
1500	0.2275	1.2439	54.06	2.376E-04	0.0904	3.193E-04	0.744
1600	0.2133	1.2541	56.36	2.642E-04	0.0948	3.545E-04	0.745
1700	0.2008	1.2630	58.61	2.919E-04	0.0992	3.913E-04	0.746
1800	0.1896	1.2708	60.83	3.208E-04	0.1035	4.295E-04	0.747
1900	0.1796	1.2778	63.01	3.508E-04	0.1077	4.692E-04	0.748
2000	0.1707	1.2841	65.16	3.818E-04	0.1118	5.104E-04	0.748

*Property values generated from NIST Database 23: REFPROP Version 7.0 (August 2002).

Table F.1G   Oxygen (O_2)*

T (K)	ρ (kg/m³)	c_p (kJ/kg·K)	μ (µPa·s)	ν (m²/s)	k (W/m·K)	α (m²/s)	Pr
100	3.995	0.9356	7.74835	1.940E-06	0.00931	2.491E-06	0.779
150	2.619	0.9198	11.34	4.329E-06	0.01399	5.807E-06	0.745
200	1.956	0.9146	14.65	7.491E-06	0.01840	1.029E-05	0.728
250	1.562	0.9150	17.71	1.134E-05	0.02260	1.581E-05	0.717
300	1.301	0.9199	20.56	1.581E-05	0.02666	2.228E-05	0.710
350	1.114	0.9291	23.23	2.085E-05	0.03067	2.962E-05	0.704
400	0.9749	0.9417	25.75	2.641E-05	0.03469	3.779E-05	0.699
450	0.8665	0.9564	28.14	3.247E-05	0.03872	4.672E-05	0.695
500	0.7798	0.9722	30.41	3.900E-05	0.04275	5.639E-05	0.692
550	0.7089	0.9880	32.58	4.597E-05	0.04676	6.677E-05	0.688
600	0.6498	1.003	34.67	5.336E-05	0.05074	7.783E-05	0.686
700	0.5569	1.031	38.62	6.935E-05	0.05850	1.019E-04	0.681
800	0.4873	1.054	42.32	8.685E-05	0.06593	1.283E-04	0.677
900	0.4332	1.074	45.82	1.058E-04	0.07300	1.569E-04	0.674
1000	0.3899	1.090	49.15	1.261E-04	0.07968	1.875E-04	0.672
1100	0.3544	1.103	52.33	1.477E-04	0.08601	2.200E-04	0.671
1200	0.3249	1.114	55.40	1.705E-04	0.09198	2.541E-04	0.671
1300	0.2999	1.123	58.36	1.946E-04	0.09762	2.897E-04	0.672
1400	0.2785	1.131	61.23	2.199E-04	0.1030	3.267E-04	0.673
1500	0.2599	1.138	64.02	2.463E-04	0.1080	3.650E-04	0.675

*Property values generated from NIST Database 23: REFPROP Version 7.0 (August 2002).

Table F.1H   Water Vapor (H_2O)*

T (K)	ρ (kg/m³)	c_p (kJ/kg·K)	μ (µPa·s)	ν (m²/s)	k (W/m·K)	α (m²/s)	Pr
373.124	0.5977	2.080	12.27	2.053E-05	0.02509	2.019E-05	1.017
400	0.5549	2.009	13.28	2.394E-05	0.02702	2.423E-05	0.988
450	0.4910	1.976	15.25	3.105E-05	0.03117	3.213E-05	0.966
500	0.4409	1.982	17.27	3.917E-05	0.03586	4.105E-05	0.954
550	0.4003	2.001	19.33	4.828E-05	0.04096	5.112E-05	0.944
600	0.3667	2.027	21.41	5.838E-05	0.04637	6.239E-05	0.936
650	0.3383	2.056	23.49	6.944E-05	0.05205	7.485E-05	0.928
700	0.3140	2.087	25.56	8.142E-05	0.05796	8.847E-05	0.920
750	0.2930	2.119	27.63	9.429E-05	0.06408	1.032E-04	0.914
800	0.2746	2.153	29.67	1.080E-04	0.07039	1.191E-04	0.907
850	0.2584	2.187	31.69	1.226E-04	0.07685	1.360E-04	0.902
900	0.2440	2.222	33.69	1.380E-04	0.08347	1.539E-04	0.897
950	0.2312	2.257	35.65	1.542E-04	0.09022	1.729E-04	0.892
1000	0.2196	2.292	37.59	1.712E-04	0.09709	1.929E-04	0.888

*Property values generated from NIST Database 23: REFPROP Version 7.0 (August 2002).

Appendix G
Thermo-Physical Properties of Selected Liquids

Table G.1 Thermo-Physical Properties of Saturated Water*

Temperature (K)	Pressure (MPa)	Liquid Density (kg/m³)	Vapor Density (kg/m³)	Liqid c_p (kJ/kg·K)	Vapor c_p (kJ/kg·K)	Liquid Viscosity (μPa·s)	Vapor Viscosity (μPa·s)	Liquid Therm. Cond. (W/m·K)	Vapor Therm. Cond. (W/m·K)	Liquid Prandtl	Vapor Prandtl	Surface Tension (N/m)	Liquid Expansion Coef. β (1/K)	Vapor Expansion Coef. β (1/K)	T (K)
273.16	0.000612	999.79	0.0048546	4.2199	1.8844	1791.2	9.2163	0.56104	0.017071	13.472	1.0173	0.075646	−0.000067965	0.0036807	273.16
280	0.000992	999.86	0.0076812	4.2014	1.8913	1433.7	9.3815	0.57404	0.017442	10.493	1.0173	0.074677	0.000043569	0.0035962	280
285	0.001389	999.47	0.010571	4.1927	1.8967	1239.3	9.509	0.58348	0.017729	8.9052	1.0173	0.073951	0.00011191	0.0035375	285
290	0.00192	998.76	0.014363	4.1869	1.9023	1084	9.6414	0.59273	0.018031	7.6573	1.0172	0.07321	0.0001721	0.0034814	290
295	0.002621	997.76	0.019281	4.1832	1.9081	957.87	9.7784	0.60169	0.018345	6.6594	1.017	0.072455	0.00022593	0.0034277	295
300	0.003537	996.51	0.02559	4.1809	1.9141	853.84	9.9195	0.61028	0.018673	5.8495	1.0168	0.071686	0.00027471	0.0033765	300
305	0.004719	995.03	0.033598	4.1798	1.9204	766.95	10.064	0.61841	0.019014	5.1837	1.0165	0.070903	0.00031942	0.0033276	305
310	0.006231	993.34	0.043663	4.1795	1.927	693.54	10.213	0.62605	0.019369	4.6301	1.0161	0.070106	0.00036081	0.0032811	310
315	0.008145	991.46	0.056195	4.1798	1.9341	630.91	10.364	0.63315	0.019736	4.1651	1.0156	0.069295	0.00039947	0.0032369	315
320	0.010546	989.39	0.071662	4.1807	1.9417	577.02	10.518	0.63971	0.020117	3.7711	1.0152	0.06847	0.00043586	0.0031951	320
325	0.013531	987.15	0.09059	4.1821	1.9499	530.29	10.675	0.64571	0.020512	3.4346	1.0147	0.067632	0.00047035	0.0031557	325
330	0.017213	984.75	0.11357	4.1838	1.9587	489.49	10.833	0.65118	0.020922	3.145	1.0143	0.066781	0.00050326	0.0031186	330
335	0.021718	982.2	0.14127	4.186	1.9684	453.64	10.994	0.65611	0.021345	2.8942	1.0139	0.065917	0.00053484	0.0030839	335
340	0.027188	979.5	0.1744	4.1885	1.979	421.97	11.157	0.66055	0.021784	2.6757	1.0136	0.06504	0.00056531	0.0030516	340
345	0.033783	976.67	0.21378	4.1913	1.9906	393.85	11.321	0.6645	0.022238	2.4842	1.0135	0.06415	0.00059485	0.0030218	345
350	0.041682	973.7	0.26029	4.1946	2.0033	368.77	11.487	0.668	0.022707	2.3156	1.0147	0.063248	0.00062362	0.0029945	350
355	0.05108	970.61	0.31487	4.1983	2.0173	346.3	11.654	0.67108	0.023193	2.1665	1.0137	0.062333	0.00065178	0.0029697	355
360	0.062194	967.39	0.37858	4.2024	2.0326	326.1	11.823	0.67376	0.023695	2.034	1.0142	0.061406	0.00067944	0.0029476	360
365	0.07526	964.05	0.45253	4.207	2.0493	307.87	11.992	0.67606	0.024213	1.9158	1.0149	0.060467	0.00070671	0.0029281	365
370	0.090535	960.59	0.53792	4.2122	2.0676	291.36	12.162	0.67802	0.02475	1.81	1.016	0.059517	0.00073371	0.0029114	370
373.15	0.10142	958.35	0.59817	4.2157	2.08	281.74	12.269	0.67909	0.025096	1.749	1.0169	0.058912	0.00075062	0.0029023	373.15
375	0.1083	957.01	0.63605	4.2178	2.0877	276.36	12.332	0.67966	0.025303	1.715	1.0175	0.058555	0.00076053	0.0028975	375
380	0.12885	953.33	0.7483	4.2241	2.1096	262.69	12.504	0.681	0.025875	1.6294	1.0194	0.057581	0.00078726	0.0028865	380
385	0.15252	949.53	0.87615	4.2309	2.1334	250.21	12.675	0.68205	0.026465	1.5521	1.0218	0.056596	0.00081398	0.0028786	385
390	0.17964	945.62	1.0212	4.2384	2.1594	238.77	12.848	0.68283	0.027074	1.4821	1.0247	0.055601	0.00084079	0.0028737	390
395	0.2106	941.61	1.185	4.2466	2.1877	228.27	13.02	0.68335	0.027701	1.4185	1.0282	0.054595	0.00086777	0.0028719	395
400	0.24577	937.49	1.3694	4.2555	2.2183	218.8	13.192	0.68364	0.028347	1.3607	1.0324	0.053578	0.00089499	0.0028735	400
405	0.28558	933.26	1.5763	4.2652	2.2514	209.68	13.365	0.68369	0.029013	1.308	1.0371	0.052551	0.00092254	0.0028784	405
410	0.33045	928.92	1.8076	4.2756	2.2871	201.43	13.538	0.68352	0.029699	1.26	1.0425	0.051514	0.0009505	0.0028868	410
415	0.38087	924.48	2.0654	4.2868	2.3254	193.78	13.711	0.68313	0.030404	1.216	1.0487	0.050468	0.00097895	0.0028987	415
420	0.4373	919.93	2.3518	4.299	2.3666	186.68	13.883	0.68253	0.031128	1.1758	1.0555	0.049411	0.001008	0.0029142	420
425	0.50025	915.27	2.6693	4.312	2.4105	180.07	14.056	0.68172	0.031873	1.139	1.063	0.048346	0.0010377	0.0029334	425
430	0.57026	910.51	3.0202	4.326	2.4573	173.91	14.228	0.6807	0.032638	1.1052	1.0712	0.047272	0.0010681	0.0029564	430
435	0.64787	905.63	3.4072	4.341	2.507	168.16	14.401	0.67948	0.033424	1.0743	1.0801	0.046189	0.0010994	0.0029834	435
440	0.73367	900.65	3.8329	4.3571	2.5597	162.77	14.573	0.67805	0.03423	1.0459	1.0898	0.045098	0.0011317	0.0030143	440
445	0.82824	895.55	4.3001	4.3743	2.6154	157.72	14.745	0.67642	0.035056	1.02	1.1001	0.043999	0.001165	0.0030495	445
450	0.9322	890.34	4.812	4.3927	2.6742	152.98	14.917	0.67459	0.035904	0.99615	1.1111	0.042891	0.0011995	0.0030889	450
455	1.0462	885.01	5.3717	4.4124	2.7362	148.52	15.089	0.67254	0.036773	0.97438	1.1228	0.041777	0.0012354	0.0031328	455

T															T
460	0.0031814	0.0012727	0.040655	1.1352	0.9545	0.037663	0.67028	15.261	144.31	2.8014	4.4334	5.9826	879.57	1.1709	460
465	0.003235	0.0013116	0.039527	1.1483	0.93639	0.038576	0.66781	15.434	140.34	2.8701	4.4559	6.6484	874	1.3069	465
470	0.0032937	0.0013522	0.038392	1.1621	0.91994	0.039512	0.66512	15.606	136.58	2.9422	4.4799	7.3727	868.31	1.4551	470
475	0.003358	0.0013948	0.037252	1.1767	0.90506	0.040471	0.66221	15.779	133.02	3.0181	4.5055	8.1598	862.49	1.616	475
480	0.0034282	0.0014396	0.036105	1.1921	0.89167	0.041455	0.65907	15.952	129.64	3.0979	4.533	9.0139	856.54	1.7905	480
485	0.0035048	0.0014867	0.034954	1.2083	0.87972	0.042464	0.65569	16.126	126.43	3.1819	4.5623	9.9397	850.45	1.9792	485
490	0.0035882	0.0015365	0.033797	1.2255	0.86914	0.043502	0.65206	16.3	123.37	3.2705	4.5937	10.942	844.22	2.1831	490
495	0.0036791	0.0015891	0.032637	1.2436	0.85989	0.044568	0.64819	16.476	120.45	3.3641	4.6274	12.027	837.84	2.4028	495
500	0.0037781	0.001645	0.031472	1.2628	0.85195	0.045666	0.64405	16.653	117.66	3.4631	4.6635	13.199	831.31	2.6392	500
505	0.0038861	0.0017045	0.030304	1.2832	0.84529	0.046799	0.63964	16.831	114.98	3.568	4.7022	14.465	824.63	2.8931	505
510	0.004004	0.001768	0.029133	1.3049	0.83991	0.047969	0.63495	17.011	112.42	3.6796	4.744	15.833	817.77	3.1655	510
515	0.0041328	0.001836	0.027959	1.3279	0.83581	0.049182	0.62997	17.193	109.95	3.7986	4.7889	17.308	810.74	3.4571	515
520	0.0042738	0.001909	0.026784	1.3524	0.83301	0.050442	0.62468	17.377	107.57	3.9257	4.8375	18.9	803.53	3.769	520
525	0.0044285	0.0019876	0.025608	1.3786	0.83154	0.051756	0.61908	17.564	105.27	4.0622	4.8901	20.617	796.13	4.1019	525
530	0.0045986	0.0020727	0.02443	1.4066	0.83143	0.05313	0.61315	17.755	103.05	4.209	4.9471	22.47	788.53	4.4569	530
535	0.0047862	0.0021652	0.023253	1.4365	0.83275	0.054575	0.60688	17.949	100.89	4.3677	5.0092	24.469	780.71	4.8349	535
540	0.0049936	0.0022659	0.022077	1.4688	0.83558	0.056102	0.60026	18.149	98.792	4.54	5.077	26.627	772.66	5.2369	540
545	0.0052239	0.0023764	0.020902	1.5031	0.84001	0.057723	0.59329	18.353	96.746	4.7277	5.1513	28.956	764.36	5.664	545
550	0.0054805	0.002498	0.01973	1.5402	0.84616	0.059456	0.58595	18.563	94.746	4.9332	5.2331	31.474	755.81	6.1172	550
555	0.0057678	0.0026326	0.018561	1.5802	0.85418	0.061321	0.57826	18.781	92.785	5.1594	5.3235	34.198	746.97	6.5976	555
560	0.0060909	0.0027826	0.017396	1.6234	0.86425	0.063341	0.57021	19.007	90.857	5.4099	5.4239	37.147	737.83	7.1062	560
565	0.0064566	0.0029508	0.016236	1.67	0.87658	0.065549	0.56181	19.242	88.956	5.6889	5.5361	40.346	728.36	7.6444	565
570	0.0068731	0.0031409	0.015082	1.7207	0.89146	0.067981	0.55308	19.489	87.074	6.002	5.6624	43.822	718.53	8.2132	570
575	0.0073509	0.0033575	0.013937	1.7758	0.90923	0.070685	0.54405	19.749	85.206	6.356	5.8055	47.607	708.3	8.814	575
580	0.007904	0.0036067	0.0128	1.8361	0.93033	0.073721	0.53474	20.024	83.342	6.7598	5.9691	51.739	697.64	9.448	580
585	0.0085503	0.0038965	0.011673	1.9025	0.95534	0.077163	0.52519	20.318	81.477	7.2252	6.1579	56.265	686.48	10.117	585
590	0.0093144	0.0042377	0.010559	1.9762	0.98504	0.081108	0.51543	20.634	79.6	7.7679	6.3784	61.242	674.78	10.821	590
595	0.01023	0.0046454	0.0094591	2.0588	1.0205	0.085682	0.50551	20.976	77.703	8.4096	6.6393	66.738	662.45	11.563	595
600	0.011345	0.0051415	0.0083756	2.1528	1.0634	0.091052	0.49546	21.35	75.773	9.1809	6.9532	72.842	649.41	12.345	600
605	0.012731	0.0057587	0.0073112	2.2619	1.116	0.097442	0.4853	21.765	73.798	10.127	7.3391	79.669	635.53	13.167	605
610	0.014496	0.0065497	0.006269	2.3917	1.1823	0.10517	0.47503	22.229	71.759	11.315	7.8268	87.369	620.65	14.033	610
615	0.016815	0.0076048	0.0052528	2.5517	1.2689	0.11468	0.46465	22.759	69.632	12.857	8.4674	96.15	604.55	14.943	615
620	0.019997	0.0090924	0.0042676	2.758	1.388	0.12666	0.4541	23.374	67.382	14.945	9.3541	106.31	586.88	15.901	620
625	0.024628	0.011355	0.0033194	3.0417	1.5635	0.14223	0.44338	24.109	64.951	17.944	10.673	118.29	567.09	16.908	625
630	0.032006	0.015162	0.0024169	3.4683	1.8459	0.16344	0.43251	25.018	62.244	22.658	12.827	132.84	544.25	17.969	630
635	0.045668	0.022437	0.0015728	4.2073	2.3528	0.19479	0.42189	26.208	59.101	31.271	16.795	151.35	516.71	19.086	635
640	0.07995	0.039706	0.00080882	5.8764	3.4542	0.25001	0.41493	27.938	55.247	52.586	25.942	177.15	481.53	20.265	640
645	0.30797	0.171	0.00017569	14.125	9.9868	0.42459	0.46136	31.348	49.357	191.32	93.35	224.45	425.05	21.515	645
647.1	—	—	0	—	—	0.19748	0.19748	39.43	39.43	—	—	322	322	22.064	647.1

*Property values generated from NIST Database 23: REFPROP Version 7.0 (August 2002).

Table G.2 Thermo–Physical Properties of Various Saturated Liquids

Table G.2A   R-134a (1,1,1,2-Tetrafluoroethane)—Saturated*

T (K)	ρ (kg/m³)	c_p (kJ/kg·K)	μ (μPa·s)	ν (m²/s)	k (W/m·K)	α (m²/s)	Pr	β (1/K)
170	1590.7	1.1838	2139.7	1.345E-06	0.14515	7.708E-08	17.45	0.001658
180	1564.2	1.1871	1479.1	9.456E-07	0.1391	7.492E-08	12.62	0.0017
190	1537.5	1.1950	1106.2	7.195E-07	0.1333	7.257E-08	9.91	0.001748
200	1510.5	1.2058	867.3	5.742E-07	0.1277	7.014E-08	8.19	0.001802
210	1483.1	1.2186	702.3	4.735E-07	0.1224	6.771E-08	6.99	0.001864
220	1455.2	1.2332	582.2	4.001E-07	0.1172	6.529E-08	6.13	0.001934
230	1426.8	1.2492	491.2	3.443E-07	0.1121	6.292E-08	5.47	0.002017
240	1397.7	1.2669	420.2	3.006E-07	0.1073	6.058E-08	4.96	0.002113
250	1367.9	1.2865	363.3	2.656E-07	0.1025	5.827E-08	4.56	0.002226
260	1337.1	1.3082	316.6	2.368E-07	0.0979	5.598E-08	4.23	0.002361
270	1305.1	1.3326	277.5	2.127E-07	0.0934	5.371E-08	3.96	0.002525
280	1271.8	1.3606	244.3	1.927E-07	0.0890	5.143E-08	3.74	0.002726
290	1236.8	1.3933	215.6	1.744E-07	0.0846	4.912E-08	3.55	0.002979
300	1199.7	1.4324	190.5	1.588E-07	0.0803	4.675E-08	3.40	0.003303
310	1159.9	1.4807	168.0	1.449E-07	0.0761	4.429E-08	3.27	0.003733
320	1116.8	1.5426	147.8	1.323E-07	0.0718	4.167E-08	3.18	0.004326
330	1069.1	1.6267	129.2	1.209E-07	0.0675	3.879E-08	3.12	0.005191
340	1015.0	1.7507	111.8	1.102E-07	0.0631	3.549E-08	3.10	0.006567
350	951.32	1.9614	95.1	9.996E-08	0.0586	3.14E-08	3.18	0.009102
360	870.11	2.4368	78.1	8.981E-08	0.0541	2.55E-08	3.52	0.015393
370	740.32	5.1048	58.0	7.829E-08	0.0518	1.37E-08	5.72	0.055237

*Property values generated from NIST Database 23: REFPROP Version 7.0 (August 2002).

Table G.2B   Engine Oil (Unused)—Saturated*

T (K)	ρ (kg/m³)	c_p (kJ/kg·K)	μ (Pa·s)	ν (m²/s)	k (W/m·K)	α (m²/s)	Pr	β (1/K)
273	899.1	1.796	3.85	0.00428	0.147	9.10E-08	47,000	0.0007
280	895.3	1.827	2.17	0.00243	0.144	8.80E-08	27,500	0.0007
290	890	1.868	0.999	0.00112	0.145	8.72E-08	12,900	0.0007
300	884.1	1.909	0.486	0.00055	0.145	8.59E-08	6,400	0.0007
310	877.9	1.951	0.253	0.000288	0.145	8.47E-08	3,400	0.0007
320	871.8	1.993	0.141	0.000161	0.143	8.23E-08	1,965	0.0007
330	865.8	2.035	0.0836	0.0000966	0.141	8.00E-08	1,205	0.0007
340	859.9	2.076	0.0531	0.0000617	0.139	7.79E-08	793	0.0007
350	853.9	2.118	0.0356	0.0000417	0.138	7.63E-08	546	0.0007
360	847.8	2.161	0.0252	0.0000297	0.138	7.53E-08	395	0.0007
370	841.8	2.206	0.0186	0.000022	0.137	7.38E-08	300	0.0007
380	836.0	2.250	0.0141	0.0000169	0.136	7.23E-08	233	0.0007
390	830.6	2.294	0.0110	0.0000133	0.135	7.09E-08	187	0.0007
400	825.1	2.337	0.00874	0.0000106	0.134	6.95E-08	152	0.0007
410	818.9	2.381	0.00698	0.00000852	0.133	6.82E-08	125	0.0007
420	812.1	2.427	0.00564	0.00000694	0.133	6.75E-08	103	0.0007
430	806.5	2.471	0.0047	0.00000583	0.132	6.62E-08	88	0.0007

*Property values from Incropera, F. P., and DeWitt, D. P., *Fundamentals of Heat and Mass Transfer*, 3rd ed., Wiley, New York, 1990.

Table G.2C   Ethylene Glycol ($C_2H_4(OH)_2$)—Saturated*

T (K)	ρ (kg/m³)	c_p (kJ/kg·K)	μ (Pa·s)	v (m²/s)	k (W/m·K)	α (m²/s)	Pr	β (1/K)
273	1130.8	2.294	0.0651	0.0000576	0.242	9.33E-08	617	0.00065
280	1125.8	2.323	0.0420	0.0000373	0.244	9.33E-08	400	0.00065
290	1118.8	2.368	0.0247	0.0000221	0.248	9.36E-08	236	0.00065
300	1114.4	2.415	0.0157	0.0000141	0.252	9.39E-08	151	0.00065
310	1103.7	2.460	0.0107	0.00000965	0.255	9.39E-08	103	0.00065
320	1096.2	2.505	0.00757	0.00000691	0.258	9.40E-08	73.5	0.00065
330	1089.5	2.549	0.00561	0.00000515	0.260	9.36E-08	55.0	0.00065
340	1083.8	2.592	0.00431	0.00000398	0.261	9.29E-08	42.8	0.00065
350	1079.0	2.637	0.00342	0.00000317	0.261	9.17E-08	34.6	0.00065
360	1074.0	2.682	0.00278	0.00000259	0.261	9.06E-08	28.6	0.00065
370	1066.7	2.728	0.00228	0.00000214	0.262	9.00E-08	23.7	0.00065
373	1058.5	2.742	0.00215	0.00000203	0.263	9.06E-08	22.4	0.00065

*Property values from Incropera, F. P., and DeWitt, D. P., *Fundamentals of Heat and Mass Transfer*, 3rd ed., Wiley, New York, 1990.

Table G.2D   Glycerin ($C_3H_5(OH)_3$)—Saturated*

T (K)	ρ (kg/m³)	c_p (kJ/kg·K)	μ (Pa·s)	v (m²/s)	k (W/m·K)	α (m²/s)	Pr	β (1/K)
273	1276.0	2.261	10.6	0.00831	0.282	9.77E-08	85,000	0.00047
280	1271.9	2.298	5.34	0.00420	0.284	9.72E-08	43,200	0.00047
290	1265.8	2.367	1.85	0.00146	0.286	9.55E-08	15,300	0.00048
300	1259.9	2.427	0.799	0.000634	0.286	9.35E-08	6,780	0.00048
310	1253.9	2.490	0.352	0.000281	0.286	9.16E-08	3,060	0.00049
320	1247.2	2.564	0.210	0.000168	0.287	8.97E-08	1,870	0.00050

*Property values from Incropera, F. P., and DeWitt, D. P., *Fundamentals of Heat and Mass Transfer*, 3rd ed., Wiley, New York, 1990.

Table G.2E   Mercury (Hg)—Saturated*

T (K)	ρ (kg/m³)	c_p (kJ/kg·K)	μ (Pa·s)	v (m²/s)	k (W/m·K)	α (m²/s)	Pr	β (1/K)
273	13,595	0.1404	0.001688	1.240E-07	8.18	4.285E-06	0.0290	0.000181
300	13,529	0.1393	0.001523	1.125E-07	8.54	4.530E-06	0.0248	0.000181
350	13,407	0.1377	0.001309	9.76E-08	9.18	4.975E-06	0.0196	0.000181
400	13,287	0.1365	0.001171	8.82E-08	9.80	5.405E-06	0.0163	0.000181
450	13,167	0.1357	0.001075	8.16E-08	10.40	5.810E-06	0.0140	0.000181
500	13,048	0.1353	0.001007	7.71E-08	10.95	6.190E-06	0.0125	0.000182
550	12,929	0.1352	0.000953	7.37E-08	11.45	6.555E-06	0.0112	0.000184
600	12,809	0.1355	0.000911	7.11E-08	11.95	6.880E-06	0.0103	0.000187

*Property values from Incropera, F. P., and DeWitt, D. P., *Fundamentals of Heat and Mass Transfer*, 3rd ed., Wiley, New York, 1990.

Appendix H
Thermo-Physical Properties of Hydrocarbon Fuels

Table H.1 Selected Properties of Hydrocarbon Fuels: Enthalpy of Formation,[a] Gibbs Function of Formation,[a] Entropy,[a] and Higher and Lower Heating Values All at 298.15 K and 1 atm; Boiling Points[b] and Latent Heat of Vaporization[c] at 1 atm; Constant-Pressure Adiabatic Flame Temperature at 1 atm;[d] Liquid Density[c]

Formula	Fuel	Molecular Weight (kg/kmol)	$\bar{h}_f^\circ$ (kJ/kmol)	$\Delta \bar{g}_f^\circ$ (kJ/kmol)	$\bar{s}^\circ$ (kJ/kmol·K)	HHV* (kJ/kg)	LHV* (kJ/kg)	Boiling Pt. (°C)	h_{fg} (kJ/kg)	$T_{ad}^\dagger$ (K)	$\rho_{liq}^\ddagger$ (kg/m³)
CH_4	Methane	16.043	−74,831	−50,794	186.188	55,528	50,016	−164	509	2226	300
C_2H_2	Acetylene	26.038	226,748	209,200	200.819	49,923	48,225	−84	—	2539	—
C_2H_4	Ethene	28.054	52,283	68,124	219.827	50,313	47,161	−103.7	—	2369	—
C_2H_6	Ethane	30.069	−84,667	−32,886	229.492	51,901	47,489	−88.6	488	2259	370
C_3H_6	Propene	42.080	20,414	62,718	266.939	48,936	45,784	−47.4	437	2334	514
C_3H_8	Propane	44.096	−103,847	−23,489	269.910	50,368	46,357	−42.1	425	2267	500
C_4H_8	1-Butene	56.107	1172	72,036	307.440	48,471	45,319	−63	391	2322	595
C_4H_{10}	n-Butane	58.123	−124,733	−15,707	310.034	49,546	45,742	−0.5	386	2270	579
C_5H_{10}	1-Pentene	70.134	−20,920	78,605	347.607	48,152	45,000	30	358	2314	641
C_5H_{12}	n-Pentane	72.150	−146,440	−8201	348.402	49,032	45,355	36.1	358	2272	626
C_6H_6	Benzene	78.113	82,927	129,658	269.199	42,277	40,579	80.1	393	2342	879
C_6H_{12}	1-Hexene	84.161	−41,673	87,027	385.974	47,955	44,803	63.4	335	2308	673
C_6H_{14}	n-Hexane	86.177	−167,193	209	386.811	48,696	45,105	69	335	2273	659
C_7H_{14}	1-Heptene	98.188	−62,132	95,563	424.383	47,817	44,665	93.6	—	2305	—
C_7H_{16}	n-Heptane	100.203	−187,820	8745	425.262	48,456	44,926	98.4	316	2274	684
C_8H_{16}	1-Octene	112.214	−82,927	104,140	462.792	47,712	44,560	121.3	—	2302	—
C_8H_{18}	n-Octane	114.230	−208,447	17,322	463.671	48,275	44,791	125.7	300	2275	703
C_9H_{18}	1-Nonene	126.241	−103,512	112,717	501.243	47,631	44,478	—	—	2300	—
C_9H_{20}	n-Nonane	128.257	−229,032	25,857	502.080	48,134	44,686	150.8	295	2276	718
$C_{10}H_{20}$	1-Decene	140.268	−124,139	121,294	539.652	47,565	44,413	170.6	—	2298	—
$C_{10}H_{22}$	n-Decane	142.284	−249,659	34,434	540.531	48,020	44,602	174.1	277	2277	730
$C_{11}H_{22}$	1-Undecene	154.295	−144,766	129,830	578.061	47,512	44,360	—	—	2296	—
$C_{11}H_{24}$	n-Undecane	156.311	−270,286	43,012	578.940	47,926	44,532	195.9	265	2277	740
$C_{12}H_{24}$	1-Dodecene	168.322	−165,352	138,407	616.471	47,468	44,316	213.4	—	2295	—
$C_{12}H_{26}$	n-Dodecane	170.337	−292,162	—	—	47,841	44,467	216.3	256	2277	749

*Based on gaseous fuel.

[†] For stoichiometric combustion with air (79% N_2, 21% O_2).

[‡] For liquids at 20°C or for gases at the boiling point of the liquefied gas.

Sources:

[a]Rossini, F. D., et al., *Selected Values of Physical and Thermodynamic Properties of Hydrocarbons and Related Compounds*, Carnegie Press, Pittsburgh, PA, 1953.

[b]Weast, R. C. (Ed.), *Handbook of Chemistry and Physics*, 56th ed., CRC Press, Cleveland, OH, 1976.

[c]Obert, E. F., *Internal Combustion Engines and Air Pollution*, Harper & Row, New York, 1973.

[d]Turns, S. R., *An Introduction to Combustion*, 2nd ed., McGraw-Hill, New York, 2000.

Table H.2 Curve-Fit Coefficients for Fuel Specific Heat and Enthalpy[a] for Reference State of Zero Enthalpy of the Elements at 298.15 K and 1 atm:

$$\bar{c}_p \text{ (kJ/kmol} \cdot \text{K)} = 4.184(a_1 + a_2\theta + a_3\theta^2 + a_4\theta^3 + a_5\theta^{-2}),$$
$$\bar{h}^\circ \text{(kJ/kmol)} = 4184(a_1\theta + a_2\theta^2/2 + a_3\theta^3/3 + a_4\theta^4/4 - a_5\theta^{-1} + a_6),$$
$$\text{where } \theta \equiv T\text{ (K)}/1000$$

Formula	Fuel	Molecular Weight	a_1	a_2	a_3	a_4	a_5	a_6	a_8^*
CH_4	Methane	16.043	−0.29149	26.327	−10.610	1.5656	0.16573	−18.331	4.300
C_3H_8	Propane	44.096	−1.4867	74.339	−39.065	8.0543	0.01219	−27.313	8.852
C_6H_{14}	Hexane	86.177	−20.777	210.48	−164.125	52.832	0.56635	−39.836	15.611
C_8H_{18}	Isooctane	114.230	−0.55313	181.62	−97.787	20.402	−0.03095	−60.751	20.232
CH_3OH	Methanol	32.040	−2.7059	44.168	−27.501	7.2193	0.20299	−48.288	5.3375
C_2H_5OH	Ethanol	46.07	6.990	39.741	−11.926	0	0	−60.214	7.6135
$C_{8.26}H_{15.5}$	Gasoline	114.8	−24.078	256.63	−201.68	64.750	0.5808	−27.562	17.792
$C_{7.76}H_{13.1}$		106.4	−22.501	227.99	−177.26	56.048	0.4845	−17.578	15.232
$C_{10.8}H_{18.7}$	Diesel	148.6	−9.1063	246.97	−143.74	32.329	0.0518	−50.128	23.514

*To obtain 0 K reference state for enthalpy, add a_8 to a_6.

[a]*Source*: From Heywood, J. B., *Internal Combustion Engine Fundamentals*, McGraw-Hill, New York, 1988, by permission of McGraw-Hill, Inc.

Table H.3 Curve-Fit Coefficients for Fuel Vapor Thermal Conductivity, Viscosity, and Specific Heat:[a]

$$\left.\begin{array}{l} k\ (\text{W/m} \cdot \text{K}) \\ \mu\ (\text{N} \cdot \text{s/m}^2) \times 10^6 \\ c_p\ (\text{J/kg} \cdot \text{K}) \end{array}\right\} = a_1 + a_2T + a_3T^2 + a_4T^3 + a_5T^4 + a_6T^5 + a_7T^6$$

Formula	Fuel	T range (K)	Property	a_1	a_2	a_3	a_4	a_5	a_6	a_7
CH_4	Methane	100–1000	k	−1.34014990E−2	3.66307060E−4	−1.82248608E−6	5.93987998E−9	−9.14055050E−12	−6.78968890E−15	−1.95048736E−18
		70–1000	μ	2.96826700E−1	3.71120100E−2	1.21829800E−5	−7.02426000E−8	7.54326900E−11	−2.72371660E−14	0
			c_p	See Table B.2						
C_3H_8	Propane	200–500	k	−1.07682209E−2	8.38590325E−5	4.22059864E−8	0	0	0	0
		270–600	μ	−3.54371100E−1	3.08009600E−2	−6.99723000E−6	0	0	0	0
			c_p	See Table B.2						
C_6H_{14}	n-Hexane	150–1000	k	1.28775700E−3	−2.00499443E−5	2.37858831E−7	−1.60944555E−10	7.71027290E−14	0	0
		270–900	μ	1.54541200E+0	1.15080900E−2	2.72216500E−5	−3.26900000E−8	1.24545900E−11	0	0
			c_p	See Table B.2						
C_7H_{16}	n-Heptane	250–1000	k	−4.60614700E−2	5.95652224E−4	−2.98893153E−6	8.44612876E−9	−1.22927E−11	9.0127E−15	−2.62961E−18
		270–580	μ	1.54009700E+0	1.09515700E−2	1.80066400E−5	−1.36379000E−8	0	0	0
		300–755	c_p	9.46260000E+1	5.86099700E+0	−1.98231320E−3	−6.88699300E−8	−1.93795260E−10	0	0
		755–1365	c_p	−7.40308000E+2	1.08935370E+2	−1.26512400E−2	9.84376300E−6	−4.32282960E−9	7.86366500E−13	0
C_8H_{18}	n-Octane	250–500	k	−4.01391940E−3	3.38796092E−5	8.19291819E−8	0	0	0	0
		300–650	μ	8.32435400E−1	1.40045000E−1	8.79376500E−6	−6.84030000E−9	0	0	0
		275–755	c_p	2.14419800E+2	5.35690500E+0	−1.17497000E−3	−6.99115500E−7	0	0	0
		755–1365	c_p	2.43596860E+3	−4.46819470E+0	−1.66843290E−2	−1.78856050E−5	8.64282020E−9	−1.6142650E−12	0
$C_{10}H_{22}$	n-Decane	250–500	k	−5.88274000E−3	3.72449646E−5	7.55109624E−8	0	0	0	0
			μ	Not available						
		300–700	c_p	2.40717800E+2	5.09965000E+0	−6.29026000E−4	−1.07155000E−6	0	0	0
		700–1365	c_p	−1.35345890E+4	9.14879000E+1	−2.20700000E−1	2.91406000E−4	−2.15307400E−7	8.38600000E−11	−1.34404000E−14
CH_3OH	Methanol	300–550	k	−2.02986750E−2	1.21910927E−4	−2.23748473E−8	0	0	0	0
		250–650	μ	1.19790000E+0	2.45028000E−2	1.86162740E−5	−1.30674820E−8	0	0	0
			c_p	See Table B.2						
C_2H_5OH	Ethanol	250–550	k	−2.46663000E−2	1.55892550E−4	−8.22954822E−8	0	0	0	0
		270–600	μ	−6.33595000E−2	3.20713470E−2	−6.25079576E−6	0	0	0	0
			c_p	See Table B.2						

[a]*Source:* Andrews, J. R., and Biblarz, O., "Temperature Dependence of Gas Properties in Polynomial Form," Naval Postgraduate School, NPS67-81-001, January 1981.

Appendix 1
Thermo-Physical Properties of Selected Solids

Table I.1 Thermo-Physical Properties of Selected Metallic Solids[a]

		Properties at 300 K				k (W/m·K) and c_p (J/kg·K) at Various Temperatures (K)									
Composition	Melting Point (K)	ρ (kg/m³)	c_p (J/kg·K)	k (W/m·K)	$\alpha \cdot 10^6$ (m²/s)	100	200	400	600	800	1000	1200	1500	2000	2500
Aluminum															
Pure	933	2702	903	237	97.1	302 / 482	237 / 798	240 / 949	231 / 1033	218 / 1146					
Alloy 2024-T6 (4.5% Cu, 1.5% Mg, 0.6% Mn)	775	2770	875	177	73.0	65 / 473	163 / 787	186 / 925	186 / 1042						
Alloy 195, Cast (4.5% Cu)		2790	883	168	68.2			174 / —	185 / —						
Beryllium	1550	1850	1825	200	59.2	990 / 203	301 / 1114	161 / 2191	126 / 2604	106 / 2823	90.8 / 3018	78.7 / 3227			
Bismuth	545	9780	122	7.86	6.59	16.5 / 112	9.69 / 120	7.04 / 127							
Boron	2573	2500	1107	27.0	9.76	190 / 128	55.5 / 600	16.8 / 1463	10.6 / 1892	9.60 / 2160	9.85 / 2338				
Cadmium	594	8650	231	96.8	48.4	203 / 198	99.3 / 222	94.7 / 242							
Chromium	2118	7160	449	93.7	29.1	159 / 192	111 / 384	90.9 / 484	80.7 / 542	71.3 / 581	65.4 / 616	61.9 / 682	57.2 / 779	49.4 / 937	
Cobalt	1769	8862	421	99.2	26.6	167 / 236	122 / 379	85.4 / 450	67.4 / 503	58.2 / 550	52.1 / 628	49.3 / 733	42.5 / 674		
Copper															
Pure	1358	8933	385	401	117	482 / 252	413 / 356	393 / 397	379 / 417	366 / 433	352 / 451	339 / 480			
Commercial bronze (90% Cu, 10% Al)	1293	8800	420	52	14		42 / 785	52 / 460	59 / 545						
Phosphor gear bronze (89% Cu, 11% Sn)	1104	8780	355	54	17		41 / —	65 / —	74 / —						
Cartridge brass (70% Cu, 30% Zn)	1188	8530	380	110	33.9	75	95 / 360	137 / 395	149 / 425						

Thermophysical Properties of Selected Metallic Solids[a] (continued)

Temperature‑dependent entries are given as k (W/m·K) / c_p (J/kg·K). Properties in the first numeric block are at 300 K unless noted.

Composition	Melting Point (K)	ρ (kg/m³)	c_p (J/kg·K)	k (W/m·K)	$\alpha\cdot10^{7}$ (m²/s)	100 K	200 K	400 K	600 K	800 K	1000 K	1200 K	1500 K
Constantan (55% Cu, 45% Ni)	1493	8920	384	23	6.71	17 / 237	19 / 362	—	—	—	—	—	—
Germanium	1211	5360	322	59.9	34.7	232 / 190	96.8 / 290	43.2 / 337	27.3 / 348	19.8 / 357	17.4 / 375	17.4 / 395	—
Gold	1336	19300	129	317	127	327 / 109	323 / 124	311 / 131	298 / 135	284 / 140	270 / 145	255 / 155	—
Iridium	2720	22500	130	147	50.3	172 / 90	153 / 122	144 / 133	138 / 138	132 / 144	126 / 153	120 / 161	111 / 172
Iron — Pure	1810	7870	447	80.2	23.1	134 / 216	94.0 / 384	69.5 / 490	54.7 / 574	43.3 / 680	32.8 / 975	28.3 / 609	32.1 / 654
Armco (99.75% pure)		7870	447	72.7	20.7	95.6 / 215	80.6 / 384	65.7 / 490	53.1 / 574	42.2 / 680	32.3 / 975	28.7 / 609	31.4 / 654
Iron — Wrought iron* (C < 0.5%)		7849	460	59	16.3	83	60	56	47	39	34	33	33
Cast iron* (C ≈ 4%)		7272	420	52	17	—	—	—	—	—	—	—	—
Carbon steels — Carbon steel* (C ≈ 0.5%)		7833	465	54	14.7	57	51	44	38	32	30	31	—
Carbon steel* (C ≈ 1.0%)		7801	473	43	11.7	43	43	39	34	30	28	29	—
Carbon steel* (C ≈ 1.5%)		7753	486	36	9.7	36	36	34	32	29	28	29	—
Carbon steels — Plain carbon (Mn ≤ 1%, Si ≤ 0.1%)		7854	434	60.5	17.7	—	—	56.7 / 487	48.0 / 559	39.2 / 685	30.0 / 1169	—	—
AISI 1010		7832	434	63.9	18.8	—	—	58.7 / 487	48.8 / 559	39.2 / 685	31.3 / 1168	—	—
Carbon–silicon (Mn ≤ 1%, 0.1% < Si ≤ 0.6%)		7817	446	51.9	14.9	—	—	49.8 / 501	44.0 / 582	37.4 / 699	29.3 / 971	—	—
Carbon–manganese–silicon (1% < Mn ≤ 1.65%, 0.1% < Si ≤ 0.6%)		8131	434	41.0	11.6	—	—	42.2 / 487	39.7 / 559	35.0 / 685	27.6 / 1090	—	—

Source: [a]Adapted from Incropera, F. P., and DeWitt, D. P., *Fundamentals of Heat and Mass Transfer*, 3rd ed., Wiley, New York, 1990, with permission. Data for wrought iron, cast iron, and various carbon steels adapted from Chapman, A. J., *Fundamentals of Heat Transfer*, Macmillan, New York, 1987.
*Properties at 293 K.

(continued)

Table I.1 (continued)

Composition	Melting Point (K)	Properties at 300 K				k (W/m·K) and c_p (J/kg·K) at Various Temperatures (K)									
		ρ (kg/m³)	c_p (J/kg·K)	k (W/m·K)	$\alpha \cdot 10^6$ (m²/s)	100	200	400	600	800	1000	1200	1500	2000	2500
Chromium (low) steels															
$\tfrac{1}{2}$Cr–$\tfrac{1}{4}$Mo–Si (0.18% C, 0.65% Cr, 0.23% Mo, 0.6% Si)		7822	444	37.7	10.9			38.2 / 492	36.7 / 575	33.3 / 688	26.9 / 969				
1 Cr–$\tfrac{1}{2}$ Mo (0.16% C, 1% Cr, 0.54% Mo, 0.39% Si)		7858	442	42.3	12.2			42.0 / 492	39.1 / 575	34.5 / 688	27.4 / 969				
1 Cr–V (0.2% C, 1.02% Cr, 0.15% V)		7836	443	48.9	14.1			46.8 / 492	42.1 / 575	36.3 / 688	28.2 / 969				
Stainless steels															
AISI 302		8055	480	15.1	3.91			17.3 / 512	20.0 / 559	22.8 / 585	25.4 / 606				
AISI 304	1670	7900	477	14.9	3.95	9.2 / 272	12.6 / 402	16.6 / 515	19.8 / 557	22.6 / 582	25.4 / 611	28.0 / 640	31.7 / 682		
AISI 316		8238	468	13.4	3.48			15.2 / 504	18.3 / 550	21.3 / 576	24.2 / 602				
AISI 347		7978	480	14.2	3.71			15.8 / 513	18.9 / 559	21.9 / 585	24.7 / 606				
Lead	601	11340	129	35.3	24.1	39.7 / 118	36.7 / 125	34.0 / 132	31.4 / 142						
Magnesium	923	1740	1024	156	87.6	169 / 649	159 / 934	153 / 1074	149 / 1170	146 / 1267					
Molybdenum	2894	10240	251	138	53.7	179 / 141	143 / 224	134 / 261	126 / 275	118 / 285	112 / 295	105 / 308	98 / 330	90 / 380	86 / 459
Nickel															
Pure	1728	8900	444	90.7	23.0	164 / 232	107 / 383	80.2 / 485	65.6 / 592	67.6 / 530	71.8 / 562	76.2 / 594	82.6 / 616		
Nichrome (80% Ni, 20% Cr)	1672	8400	420	12	3.4			14 / 480	16 / 525	21 / 545					

Material																
Inconel X-750 (73% Ni, 15% Cr, 6.7% Fe)	1665	8510	439	11.7	3.1	8.7 / —	10.3 / 372	13.5 / 473	17.0 / 510	20.5 / 546	24.0 / 626	27.6 / —	33.0 / —			
Niobium	2741	8570	265	53.7	23.6	55.2 / 188	52.6 / 249	55.2 / 274	58.2 / 283	61.3 / 292	64.4 / 301	67.5 / 310	72.1 / 324	79.1 / 347		
Palladium	1827	12020	244	71.8	24.5	76.5 / 168	71.6 / 227	73.6 / 251	79.7 / 261	86.9 / 271	94.2 / 281	102 / 291	110 / 307	110 / 349		
Platinum Pure	2045	21450	133	71.6	25.1	77.5 / 100	72.6 / 125	71.8 / 136	73.2 / 141	75.6 / 146	78.7 / 152	82.6 / 157	89.5 / 165	99.4 / 179		
Alloy 60Pt–40Rh (60% Pt, 40% Rh)	1800	16630	162	47	17.4	52 / —	59 / —	65 / —	69 / —	73 / —	76 / —					
Rhenium	3453	21100	136	47.9	16.7	58.9 / 97	51.0 / 127	46.1 / 139	44.2 / 145	44.1 / 151	44.6 / 156	45.7 / 162	47.8 / 171	51.9 / 186		
Rhodium	2236	12450	243	150	49.6	186 / 147	154 / 220	146 / 253	136 / 274	127 / 293	121 / 311	116 / 327	110 / 349	112 / 376		
Silicon	1685	2330	712	148	89.2	884 / 259	264 / 556	98.9 / 790	61.9 / 867	42.2 / 913	31.2 / 946	25.7 / 967	22.7 / 992			
Silver	1235	10500	235	429	174	444 / 187	430 / 225	425 / 239	412 / 250	396 / 262	379 / 277	361 / 292				
Tantalum	3269	16600	140	57.5	24.7	59.2 / 110	57.5 / 133	57.8 / 144	58.6 / 146	59.4 / 149	60.2 / 152	61.0 / 155	62.2 / 160	64.1 / 172	65.6 / 189	
Thorium	2023	11700	118	54.0	39.1	59.8 / 99	54.6 / 112	54.5 / 124	55.8 / 134	56.9 / 145	56.9 / 156	58.7 / 167				
Tin	505	7310	227	66.6	40.1	85.2 / 188	73.3 / 215	62.2 / 243								
Titanium	1953	4500	522	21.9	9.32	30.5 / 300	24.5 / 465	20.4 / 551	19.4 / 591	19.7 / 633	20.7 / 675	22.0 / 620	24.5 / 686			
Tungsten	3660	19300	132	174	68.3	208 / 87	186 / 122	159 / 137	137 / 142	125 / 145	118 / 148	113 / 152	110 / 157	107 / 157	100 / 167	95 / 176
Uranium	1406	19070	116	27.6	12.5	21.7 / 94	25.1 / 108	29.6 / 125	34.0 / 146	38.8 / 176	43.9 / 180	49.0 / 161				

(continued)

Table I.1 (continued)

Composition	Melting Point (K)	Properties at 300 K				k (W/m·K) and c_p (J/kg·K) at Various Temperatures (K)									
		ρ (kg/m³)	c_p (J/kg·K)	k (W/m·K)	$\alpha \cdot 10^6$ (m²/s)	100	200	400	600	800	1000	1200	1500	2000	2500
Vanadium	2192	6100	489	30.7	10.3	35.8 / 258	31.3 / 430	31.3 / 515	33.3 / 540	35.7 / 563	38.2 / 597	40.8 / 645	44.6 / 714	50.9 / 867	
Zinc	693	7140	389	116	41.8	117 / 297	118 / 367	111 / 402	103 / 436						
Zirconium	2125	6570	278	22.7	12.4	33.2 / 205	25.2 / 264	21.6 / 300	20.7 / 322	21.6 / 342	23.7 / 362	26.0 / 344	28.8 / 344	33.0 / 344	

Table I.2 Thermo-Physical Properties of Selected Nonmetallic Solids[a]

Composition	Melting Point (K)	Properties at 300 K ρ (kg/m³)	c_p (J/kg·K)	k (W/m·K)	$\alpha \cdot 10^6$ (m²/s)	k (W/m·K) and c_p (J/kg·K) at Various Temperatures (K) 100	200	400	600	800	1000	1200	1500	2000	2500
Aluminum oxide, sapphire	2323	3970	765	46	15.1	450 / —	82 / —	32.4 / 940	18.9 / 1110	13.0 / 1180	10.5 / 1225				
Aluminum oxide, polycrystalline	2323	3970	765	36.0	11.9	133 / —	55 / —	26.4 / 940	15.8 / 1110	10.4 / 1180	7.85 / 1225	6.55 / —	5.66 / —	6.00 / —	
Beryllium oxide	2725	3000	1030	272	88.0			196 / 1350	111 / 1690	70 / 1865	47 / 1975	33 / 2055	21.5 / 2145	15 / 2750	
Boron	2573	2500	1105	27.6	9.99	190 / —	52.5 / —	18.7 / 1490	11.3 / 1880	8.1 / 2135	6.3 / 2350	5.2 / 2555			
Boron fiber epoxy (30% vol) composite	590	2080													
k, ∥ to fibers				2.29		2.10	2.23	2.28							
k, ⊥ to fibers				0.59		0.37	0.49	0.60							
c_p			1122			364	757	1431							
Carbon															
Amorphous	1500	1950	—	1.60	—	0.67 / —	1.18 / —	1.89 / —	2.19 / —	2.37 / —	2.53 / —	2.84 / —	3.48 / —		
Diamond, type IIa insulator	—	3500	509	2300	—	10000 / 21	4000 / 194	1540 / 853							
Graphite, pyrolytic	2273	2210													
k, ∥ to layers				1950		4970	3230	1390	892	667	534	448	357	262	
k, ⊥ to layers				5.70		16.8	9.23	4.09	2.68	2.01	1.60	1.34	1.08	0.81	
c_p			709			136	411	992	1406	1650	1793	1890	1974	2043	
Graphite fiber epoxy (25% vol) composite	450	1400													
k, heat flow ∥ to fibers				11.1		5.7	8.7	13.0							
k, heat flow ⊥ to fibers				0.87		0.46	0.68	1.1							
c_p			935			337	642	1216							

(continued)

Table I.2 (continued)

Composition	Melting Point (K)	Properties at 300 K ρ (kg/m³)	c_p (J/kg·K)	k (W/m·K)	$\alpha \cdot 10^6$ (m²/s)	k (W/m·K) and c_p (J/kg·K) at Various Temperatures (K) 100	200	400	600	800	1000	1200	1500	2000	2500
Pyroceram, Corning 9606	1623	2600	808	3.98	1.89	5.25 —	4.78 —	3.64 908	3.28 1038	3.08 1122	2.96 1197	2.87 1264	2.79 1498		
Silicon carbide	3100	3160	675	490	230			— 880	— 1050	— 1135	87 1195	58 1243	30 1310		
Silicon dioxide, crystalline (quartz) k, ∥ to c axis k, ⊥ to c axis c_p	1883	2650	745	10.4 6.21		39 20.8 —	16.4 9.5 —	7.6 4.70 885	5.0 3.4 1075	4.2 3.1 1250					
Silicon dioxide, polycrystalline (fused silica)	1883	2220	745	1.38	0.834	0.69 —	1.14 —	1.51 905	1.75 1040	2.17 1105	2.87 1155	4.00 1195			
Silicon nitride	2173	2400	691	16.0	9.65	— —	— 578	13.9 778	11.3 937	9.88 1063	8.76 1155	8.00 1226	7.16 1306	6.20 1377	
Sulfur	392	2070	708	0.206	0.141	0.165 403	0.185 606								
Thorium dioxide	3573	9110	235	13	6.1			10.2 255	6.6 274	4.7 285	3.68 295	3.12 303	2.73 315	2.5 330	
Titanium dioxide, polycrystalline	2133	4157	710	8.4	2.8			7.01 805	5.02 880	3.94 910	3.46 930	3.28 945			

Source: [a]Adapted from Incropera, F. P., and Dewitt, D. P., *Fundamentals of Heat and Mass Transfer*, 3rd ed., Wiley, New York, 1990, with permission.

Table I.3 Thermo-Physical Properties of Common Materials[a]

Description/ Composition	Temperature (K)	Density, ρ (kg/m³)	Thermal Conductivity, k (W/m·K)	Specific Heat, c_p (J/kg·K)
Asphalt	300	2115	0.062	920
Bakelite	300	1300	1.4	1465
Brick, refractory				
Carborundum	872	—	18.5	—
	1672	—	11.0	—
Chrome brick	473	3010	2.3	835
	823		2.5	
	1173		2.0	
Diatomaceous	478	—	0.25	
silica, fired	1145	—	0.30	—
Fire clay, burnt 1600 K	773	2050	1.0	960
	1073	—	1.1	
	1373	—	1.1	
Fire clay, burnt 1725 K	773	2325	1.3	960
	1073		1.4	
	1373		1.4	
Fire clay brick	478	2645	1.0	960
	922		1.5	
	1478		1.8	
Magnesite	478	—	3.8	1130
	922	—	2.8	
	1478		1.9	
Clay	300	1460	1.3	880
Coal, anthracite	300	1350	0.26	1260
Concrete (stone mix)	300	2300	1.4	880
Cotton	300	80	0.06	1300
Foodstuffs				
Banana (75.7% water content)	300	980	0.481	3350
Apple, red (75% water content)	300	840	0.513	3600
Cake, batter	300	720	0.223	—
Cake, fully baked	300	280	0.121	—
Chicken meat, white	198	—	1.60	—
(74.4% water content)	233	—	1.49	
	253		1.35	
	263		1.20	
	273		0.476	
	283		0.480	
	293		0.489	
Glass				
Plate (soda lime)	300	2500	1.4	750
Pyrex	300	2225	1.4	835

(continued)

Table I.3 (continued)

Description/ Composition	Temperature (K)	Density, ρ (kg/m^3)	Thermal Conductivity, k (W/m·K)	Specific Heat, c_p (J/kg·K)
Ice	273	920	1.88	2040
	253	—	2.03	1945
Leather (sole)	300	998	0.159	—
Paper	300	930	0.180	1340
Paraffin	300	900	0.240	2890
Rock				
Granite, Barre	300	2630	2.79	775
Limestone, Salem	300	2320	2.15	810
Marble, Halston	300	2680	2.80	830
Quartzite, Sioux	300	2640	5.38	1105
Sandstone, Berea	300	2150	2.90	745
Rubber, vulcanized				
Soft	300	1100	0.13	2010
Hard	300	1190	0.16	—
Sand	300	1515	0.27	800
Soil	300	2050	0.52	1840
Snow	273	110	0.049	—
		500	0.190	—
Teflon	300	2200	0.35	—
	400		0.45	—
Tissue, human				
Skin	300	—	0.37	—
Fat layer (adipose)	300	—	0.2	—
Muscle	300	—	0.41	—
Wood, cross grain				
Balsa	300	140	0.055	—
Cypress	300	465	0.097	—
Fir	300	415	0.11	2720
Oak	300	545	0.17	2385
Yellow pine	300	640	0.15	2805
White pine	300	435	0.11	—
Wood, radial				
Oak	300	545	0.19	2385
Fir	300	420	0.14	2720

Source: [a]Adapted from Incropera, F. P., and DeWitt, D. P., *Fundamentals of Heat and Mass Transfer,* 3rd ed., Wiley, New York, 1990, with permission.

Table I.4   Thermo-Physical Properties of Structural Building Materials[a]

Description/ Composition	Typical Properties at 300 K		
	Density, ρ (kg/m³)	Thermal Conductivity, k (W/m·K)	Specific Heat, c_p (J/kg·K)
Building boards			
Asbestos–cement board	1920	0.58	—
Gypsum or plaster board	800	0.17	—
Plywood	545	0.12	1215
Sheathing, regular density	290	0.055	1300
Acoustic tile	290	0.058	1340
Hardboard, siding	640	0.094	1170
Hardboard, high density	1010	0.15	1380
Particle board, low density	590	0.078	1300
Particle board, high density	1000	0.170	1300
Woods			
Hardwoods (oak, maple)	720	0.16	1255
Softwoods (fir, pine)	510	0.12	1380
Masonry materials			
Cement mortar	1860	0.72	780
Brick, common	1920	0.72	835
Brick, face	2083	1.3	—
Clay tile, hollow			
1 cell deep, 10 cm thick	—	0.52	—
3 cells deep, 30 cm thick	—	0.69	—
Concrete block, 3 oval cores			
sand/gravel, 20 cm thick	—	1.0	—
cinder aggregate, 20 cm thick	—	0.67	—
Concrete block, rectangular core			
2 cores, 20 cm thick, 16 kg	—	1.1	—
same with filled cores	—	0.60	—
Plastering materials			
Cement plaster, sand aggregate	1860	0.72	—
Gypsum plaster, sand aggregate	1680	0.22	1085
Gypsum plaster, vermiculite aggregate	720	0.25	—
Blanket and batt			
Glass fiber, paper faced	16	0.046	—
	28	0.038	—
	40	0.035	—
Glass fiber, coated; duct liner	32	0.038	835
Board and slab			
Cellular glass	145	0.058	1000
Glass fiber, organic bonded	105	0.036	795
Polystyrene, expanded			
extruded (R-12)	55	0.027	1210
molded beads	16	0.040	1210
Mineral fiberboard; roofing material	265	0.049	—
Wood, shredded/cemented	350	0.087	1590
Cork	120	0.039	1800

(*continued*)

Table I.4   (continued)

Description/ Composition	Typical Properties at 300 K		
	Density, ρ (kg/m³)	Thermal Conductivity, k (W/m·K)	Specific Heat, c_p (J/kg·K)
Loose fill			
Cork, granulated	160	0.045	—
Diatomaceous silica, coarse	350	0.069	—
powder	400	0.091	—
Diatomaceous silica, fine powder	200	0.052	—
	275	0.061	—
Glass, fiber, poured or blown	16	0.043	835
Vermiculite, flakes	80	0.068	835
	160	0.063	1000
Formed/foamed-in-place			
Mineral wool granules with asbestos/inorganic binders, sprayed	190	0.046	—
Polyvinyl acetate cork mastic; sprayed or troweled	—	0.100	—
Urethane, two-part mixture; rigid foam	70	0.026	1045
Reflective			
Aluminum foil separating fluffy glass mats; 10–12 layers; evacuated; for cryogenic applications (150 K)	40	0.00016	—
Aluminum foil and glass paper laminate; 75–150 layers; evacuated; for cryogenic application (150 K)	120	0.000017	—
Typical silica powder, evacuated	160	0.0017	—

Source: [a]Adapted from Incropera, F. P., and DeWitt, D. P., *Fundamentals of Heat and Mass Transfer,* 3rd ed., Wiley, New York, 1990, with permission.

Table I.5 Thermo-Physical Properties of Industrial Insulation[a]

Typical Thermal Conductivity k (W/m · K) at Various Temperatures (K)

Description/Composition	Max Service Temp (K)	Typical Density (kg/m³)	200	215	230	240	255	270	285	300	310	365	420	530	645	750
Blankets																
Blanket, mineral fiber, metal reinforced	920	96–192									0.038	0.046	0.056	0.078		
	815	40–96									0.035	0.045	0.058	0.088		
Blanket, mineral fiber, glass; fine fiber, organic bonded	450	10				0.036	0.038	0.040	0.043	0.048	0.052	0.076				
		12				0.035	0.036	0.039	0.042	0.046	0.049	0.069				
		16				0.033	0.035	0.036	0.039	0.042	0.046	0.062				
		24				0.030	0.032	0.033	0.036	0.039	0.040	0.053				
		32				0.029	0.030	0.032	0.033	0.036	0.038	0.048				
		48				0.027	0.029	0.030	0.032	0.033	0.035	0.045				
Blanket, alumina–silica fiber	1530	48												0.071	0.105	0.150
		64												0.059	0.087	0.125
		96												0.052	0.076	0.100
		128												0.049	0.068	0.091
Felt, semirigid; organic bonded	480	50–125	0.023	0.025	0.026	0.027	0.029	0.030	0.032	0.033	0.035	0.051	0.063			
	730	50						0.035	0.036	0.038	0.039	0.051	0.079			
Felt, laminated; no binder	920	120											0.051	0.065	0.087	
Blocks, boards, and pipe insulations																
Asbestos paper, laminated and corrugated																
4-ply	420	190								0.078	0.082	0.098				
6-ply	420	255								0.071	0.074	0.085				
8-ply	420	300								0.068	0.071	0.082				
Magnesia, 85%	590	185									0.051	0.055	0.061			
Calcium silicate	920	190									0.055	0.059	0.063	0.075	0.089	0.104
Cellular glass	700	145			0.046	0.048	0.051	0.052	0.055	0.058	0.062	0.069	0.079			
Diatomaceous silica	1145	345												0.092	0.098	0.104
	1310	385												0.101	0.100	0.115

(continued)

Table I.5 (continued)

Description/Composition	Max Service Temp (K)	Typical Density (kg/m³)	Typical Thermal Conductivity k (W/m · K) at Various Temperatures (K)													
			200	215	230	240	255	270	285	300	310	365	420	530	645	750
Polystyrene, rigid																
Extruded (R-12)	350	56	0.023	0.023	0.022	0.023	0.023	0.025	0.026	0.027	0.029					
Extruded (R-12)	350	35	0.023	0.023	0.023	0.025	0.025	0.026	0.027	0.029						
Molded beads	350	16	0.026	0.029	0.030	0.033	0.035	0.036	0.038	0.040						
Rubber, rigid foamed	340	70						0.029	0.030	0.032	0.033					
Insulating cement																
Mineral fiber (rock, slag, or glass)																
with clay binder	1255	430									0.071	0.079	0.088	0.105	0.123	
with hydraulic setting binder	922	560									0.108	0.115	0.123	0.137		
Loose fill																
Cellulose, wood, or paper pulp	—	45							0.038	0.039	0.042					
Perlite, expanded	—	105	0.036	0.039	0.042	0.043	0.046	0.049	0.051	0.053	0.056					
Vermiculite, expanded	—	122			0.056	0.058	0.061	0.063	0.065	0.068	0.071					
Vermiculite, expanded	—	80			0.049	0.051	0.055	0.058	0.061	0.063	0.066					

Source: [a]Adapted from Incropera, F. P., and DeWitt, D. P., *Fundamentals of Heat and Mass Transfer*, 3rd ed., Wiley, New York, 1990, with permission.

Appendix J
Radiation Properties of Selected Materials and Substances

Table J.1   Total, Normal (n), or Hemispherical (h) Emissivity of Selected Surfaces: Metallic Solids and Their Oxides[a]

Description/Composition		Emissivity at Various Temperatures (K)										
		100	200	300	400	600	800	1000	1200	1500	2000	2500
Aluminum												
Highly polished, film	(h)	0.02	0.03	0.04	0.05	0.06						
Foil, bright	(h)	0.06	0.06	0.07								
Anodized	(h)			0.82	0.76							
Chromium												
Polished or plated	(n)	0.05	0.07	0.10	0.12	0.14						
Copper												
Highly polished	(h)			0.03	0.03	0.04	0.04	0.04				
Stably oxidized	(h)					0.50	0.58	0.80				
Gold												
Highly polished or film	(h)	0.01	0.02	0.03	0.03	0.04	0.05	0.06				
Foil, bright	(h)	0.06	0.07	0.07								
Molybdenum												
Polished	(h)					0.06	0.08	0.10	0.12	0.15	0.21	0.26
Shot-blasted, rough	(h)					0.25	0.28	0.31	0.35	0.42		
Stably oxidized	(h)					0.80	0.82					
Nickel												
Polished	(h)					0.09	0.11	0.14	0.17			
Stably oxidized	(h)					0.40	0.49	0.57				
Platinum												
Polished	(h)						0.10	0.13	0.15	0.18		
Silver												
Polished	(h)			0.02	0.02	0.03	0.05	0.08				
Stainless steels												
Typical, polished	(n)			0.17	0.17	0.19	0.23	0.30				
Typical, cleaned	(n)			0.22	0.22	0.24	0.28	0.35				
Typical, lightly oxidized	(n)						0.33	0.40				
Typical, highly oxidized	(n)						0.67	0.70	0.76			
AISI 347, stably oxidized	(n)					0.87	0.88	0.89	0.90			
Tantalum												
Polished	(h)								0.11	0.17	0.23	0.28
Tungsten												
Polished	(h)							0.10	0.13	0.18	0.25	0.29

Source: [a]Adapted from Incropera, F. P., and DeWitt, D. P., *Fundamentals of Heat and Mass Transfer*, 3rd ed., Wiley, New York, 1990, with permission.

Table J.2 Total, Normal (n), or Hemispherical (h) Emissivity of Selected Surfaces: Nonmetallic Substances

Description/Composition		Temperature (K)	Emissivity ε
Aluminum oxide	(n)	600	0.69
		1000	0.55
		1500	0.41
Asphalt pavement	(h)	300	0.85–0.93
Building materials			
Asbestos sheet	(h)	300	0.93–0.96
Brick, red	(h)	300	0.93–0.96
Gypsum or plaster board	(h)	300	0.90–0.92
Wood	(h)	300	0.82–0.92
Cloth	(h)	300	0.75–0.90
Concrete	(h)	300	0.88–0.93
Glass, window	(h)	300	0.90–0.95
Ice	(h)	273	0.95–0.98
Paints			
Black (Parsons)	(h)	300	0.98
White, acrylic	(h)	300	0.90
White, zinc oxide	(h)	300	0.92
Paper, white	(h)	300	0.92–0.97
Pyrex	(n)	300	0.82
		600	0.80
		1000	0.71
		1200	0.62
Pyroceram	(n)	300	0.85
		600	0.78
		1000	0.69
		1500	0.57
Refractories (furnace liners)			
Alumina brick	(n)	800	0.40
		1000	0.33
		1400	0.28
		1600	0.33
Magnesia brick	(n)	800	0.45
		1000	0.36
		1400	0.31
		1600	0.40
Kaolin insulating brick	(n)	800	0.70
		1200	0.57
		1400	0.47
		1600	0.53
Sand	(h)	300	0.90
Silicon carbide	(n)	600	0.87
		1000	0.87
		1500	0.85
Skin	(h)	300	0.95
Snow	(h)	273	0.82–0.90
Soil	(h)	300	0.93–0.96
Rocks	(h)	300	0.88–0.95
Teflon	(h)	300	0.85
		400	0.87
		500	0.92
Vegetation	(h)	300	0.92–0.96
Water	(h)	300	0.96

Source: [a]Adapted from Incropera, F. P., and DeWitt, D. P., *Fundamentals of Heat and Mass Transfer*, 3rd ed., Wiley, New York, 1990, with permission.

Appendix K
Mach Number Relationships for Compressible Flow

Table K.1 One-Dimensional, Isentropic, Variable-Area Flow of Air with Constant Properties ($\gamma = 1.4$)

Ma	$\dfrac{A}{A^*}$	$\dfrac{P}{P_0}$	$\dfrac{\rho}{\rho_0}$	$\dfrac{T}{T_0}$
0	∞	1.00000	1.00000	1.00000
0.10	5.8218	0.99303	0.99502	0.99800
0.20	2.9635	0.97250	0.98027	0.99206
0.30	2.0351	0.93947	0.95638	0.98232
0.40	1.5901	0.89562	0.92428	0.96899
0.50	1.3398	0.84302	0.88517	0.95238
0.60	1.1882	0.78400	0.84045	0.93284
0.70	1.09437	0.72092	0.79158	0.91075
0.80	1.03823	0.65602	0.74000	0.88652
0.90	1.00886	0.59126	0.68704	0.86058
1.00	1.00000	0.52828	0.63394	0.83333
1.10	1.00793	0.46835	0.58169	0.80515
1.20	1.03044	0.41238	0.53114	0.77640
1.30	1.06631	0.36092	0.48291	0.74738
1.40	1.1149	0.31424	0.43742	0.71839
1.50	1.1762	0.27240	0.39498	0.68965
1.60	1.2502	0.23527	0.35573	0.66138
1.70	1.3376	0.20259	0.31969	0.63372
1.80	1.4390	0.17404	0.28682	0.60680
1.90	1.5552	0.14924	0.25699	0.58072
2.00	1.6875	0.12780	0.23005	0.55556
2.10	1.8369	0.10935	0.20580	0.53135
2.20	2.0050	0.09352	0.18405	0.50813
2.30	2.1931	0.07997	0.16458	0.48591
2.40	2.4031	0.06840	0.14720	0.46468
2.50	2.6367	0.05853	0.13169	0.44444
2.60	2.8960	0.05012	0.11787	0.42517
2.70	3.1830	0.04295	0.10557	0.40684
2.80	3.5001	0.03685	0.09462	0.38941
2.90	3.8498	0.03165	0.08489	0.37286
3.00	4.2346	0.02722	0.07623	0.35714
3.50	6.7896	0.01311	0.04523	0.28986
4.00	10.719	0.00658	0.02766	0.23810
4.50	16.562	0.00346	0.01745	0.19802
5.00	25.000	$189(10)^{-5}$	0.01134	0.16667
6.00	53.180	$633(10)^{-6}$	0.00519	0.12195
7.00	104.143	$242(10)^{-6}$	0.00261	0.09259
8.00	190.109	$102(10)^{-6}$	0.00141	0.07246
9.00	327.189	$474(10)^{-7}$	0.000815	0.05814
10.00	535.938	$236(10)^{-7}$	0.000495	0.04762
∞	∞	0	0	0

Table K.2   One-Dimensional Normal-Shock Functions for Air with Constant Properties ($\gamma = 1.4$)

Ma_x	Ma_y	$\dfrac{P_y}{P_x}$	$\dfrac{\rho_y}{\rho_x}$	$\dfrac{T_y}{T_x}$	$\dfrac{P_{0y}}{P_{0x}}$	$\dfrac{P_{0y}}{P_x}$
1.00	1.00000	1.0000	1.0000	1.0000	1.00000	1.8929
1.10	0.91177	1.2450	1.1691	1.0649	0.99892	2.1328
1.20	0.84217	1.5133	1.3416	1.1280	0.99280	2.4075
1.30	0.78596	1.8050	1.5157	1.1909	0.97935	2.7135
1.40	0.73971	2.1200	1.6896	1.2547	0.95819	3.0493
1.50	0.70109	2.4583	1.8621	1.3202	0.92978	3.4133
1.60	0.66844	2.8201	2.0317	1.3880	0.89520	3.8049
1.70	0.64055	3.2050	2.1977	1.4583	0.85573	4.2238
1.80	0.61650	3.6133	2.3592	1.5316	0.81268	4.6695
1.90	0.59562	4.0450	2.5157	1.6079	0.76735	5.1417
2.00	0.57735	4.5000	2.6666	1.6875	0.72088	5.6405
2.10	0.56128	4.9784	2.8119	1.7704	0.67422	6.1655
2.20	0.54706	5.4800	2.9512	1.8569	0.62812	6.7163
2.30	0.53441	6.0050	3.0846	1.9468	0.58331	7.2937
2.40	0.52312	6.5533	3.2119	2.0403	0.54015	7.8969
2.50	0.51299	7.1250	3.3333	2.1375	0.49902	8.5262
2.60	0.50387	7.7200	3.4489	2.2383	0.46012	9.1813
2.70	0.49563	8.3383	3.5590	2.3429	0.42359	9.8625
2.80	0.48817	8.9800	3.6635	2.4512	0.38946	10.569
2.90	0.48138	9.6450	3.7629	2.5632	0.35773	11.302
3.00	0.47519	10.333	3.8571	2.6790	0.32834	12.061
4.00	0.43496	18.500	4.5714	4.0469	0.13876	21.068
5.00	0.41523	29.000	5.0000	5.8000	0.06172	32.654
10.00	0.38757	116.50	5.7143	20.388	0.00304	129.217
∞	0.37796	∞	6.000	∞	0	∞

Appendix L
Psychrometry Chart

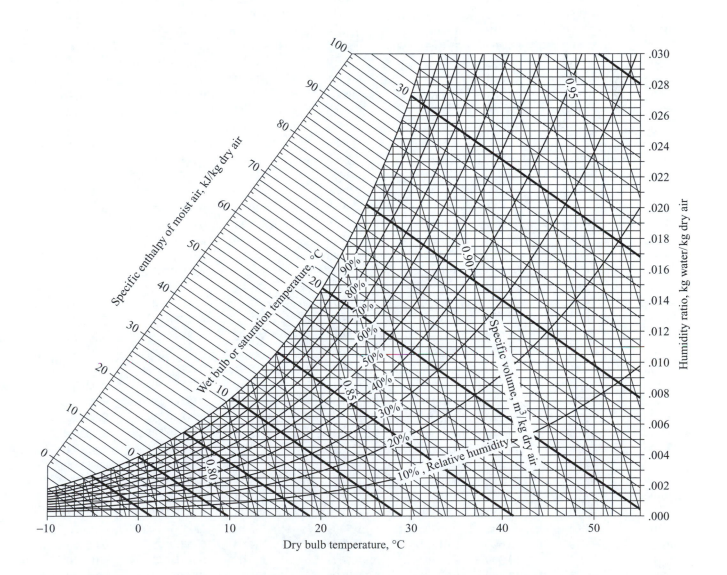

FIGURE L.1

Psychrometric chart in SI units (P = 1 atm). **Adapted with permission from Z. Zhang and M. B. Pate, "A Methodology for Implementing a Psychrometric Chart in a Computer Graphics System,"** *ASHRAE Transactions***, Vol. 94, Pt. 1, 1988.**

Appendix M
Properties of the Atmosphere at High Altitude

Table M.1 Properties of the Atmosphere at High Altitude*

Altitude (m)	Temperature (°C)	Pressure (kPa)	Density (kg/m^3)	Acceleration due to Gravity (m/s^2)	Speed of Sound (m/s)	Viscosity (N·s/m^2)	Thermal Conductivity (W/m·K)
0	15.0	101.32	1.2250	9.8066	340.29	0.000017894	0.025326
200	13.7	98.945	1.2017	9.8060	339.53	0.000017831	0.025224
400	12.4	96.611	1.1786	9.8054	338.76	0.000017768	0.025121
600	11.1	94.322	1.1560	9.8048	337.98	0.000017705	0.025019
800	9.80	92.078	1.1337	9.8042	337.21	0.000017642	0.024916
1000	8.50	89.876	1.1117	9.8036	336.43	0.000017579	0.024813
1200	7.20	87.718	1.0900	9.8029	335.66	0.000017515	0.024710
1400	5.90	85.602	1.0687	9.8023	334.88	0.000017451	0.024607
1600	4.60	83.528	1.0476	9.8017	334.10	0.000017388	0.024504
1800	3.30	81.494	1.0269	9.8011	333.32	0.000017324	0.024401
2000	2.00	79.501	1.0066	9.8005	332.53	0.000017260	0.024298
2200	0.70	77.548	0.98648	9.7999	331.75	0.000017196	0.024194
2400	0.59	75.634	0.96672	9.7992	330.96	0.000017131	0.024091
2600	−1.89	73.759	0.94726	9.7986	330.17	0.000017067	0.023987
2800	−3.19	71.921	0.92811	9.7980	329.38	0.000017002	0.023883
3000	−4.49	70.121	0.90925	9.7974	328.58	0.000016938	0.023779
3200	−5.79	68.358	0.89069	9.7968	327.79	0.000016873	0.023675
3400	−7.09	66.631	0.87243	9.7962	326.99	0.000016808	0.023571
3600	−8.39	64.939	0.85445	9.7956	326.19	0.000016743	0.023466
3800	−9.69	63.282	0.83676	9.7949	325.39	0.000016677	0.023362
4000	−11.0	61.660	0.81935	9.7943	324.59	0.000016612	0.023257
4200	−12.3	60.072	0.80222	9.7937	323.78	0.000016546	0.023152
4400	−13.6	58.518	0.78536	9.7931	322.98	0.000016481	0.023047
4600	−14.9	56.996	0.76878	9.7925	322.17	0.000016415	0.022942
4800	−16.2	55.506	0.75247	9.7919	321.36	0.000016349	0.022837
5000	−17.5	54.048	0.73643	9.7912	320.55	0.000016282	0.022732
5200	−18.8	52.622	0.72065	9.7906	319.73	0.000016216	0.022627
5400	−20.1	51.226	0.70513	9.7900	318.91	0.000016150	0.022521
5600	−21.4	49.860	0.68987	9.7894	318.10	0.000016083	0.022415
5800	−22.7	48.524	0.67486	9.7888	317.27	0.000016016	0.022310
6000	−24.0	47.218	0.66011	9.7882	316.45	0.000015949	0.022204
6200	−25.3	45.940	0.64561	9.7875	315.63	0.000015882	0.022098
6400	−26.6	44.690	0.63135	9.7869	314.80	0.000015815	0.021991
6600	−27.9	43.468	0.61733	9.7863	313.97	0.000015748	0.021885
6800	−29.2	42.273	0.60356	9.7857	313.14	0.000015680	0.021779
7000	−30.5	41.105	0.59002	9.7851	312.31	0.000015612	0.021672
8000	−36.9	35.652	0.52579	9.7820	308.11	0.000015271	0.021138
9000	−43.4	30.801	0.46706	9.7789	303.85	0.000014926	0.020600
10000	−49.9	26.500	0.41351	9.7759	299.53	0.000014577	0.020059
12000	−56.5	19.399	0.31194	9.7697	295.07	0.000014216	0.019505
14000	−56.5	14.170	0.22786	9.7636	295.07	0.000014216	0.019505
16000	−56.5	10.353	0.16647	9.7575	295.07	0.000014216	0.019505
18000	−56.5	7.5652	0.12165	9.7513	295.07	0.000014216	0.019505
20000	−56.5	5.5293	0.088910	9.7452	295.07	0.000014216	0.019505

*United States Committee on Extension to the Standard Atmosphere, "U.S. Standard Atmosphere, 1976," National Oceanic and Atmospheric Administration, National Aeronautics and Space Administration, United States Air Force, Washington D.C., 1976.

U.S. Customary Units, Appendices AE–ME

Appendix AE
Thermodynamic Properties of Refrigerant-134a

Table AE.1 Saturation Properties of Refrigerant R-134a: Temperature Increments

Temp., °F	Press., P_{sat}, psia	Specific Volume, ft³/lbm		Internal Energy, Btu/lbm		Enthalpy, Btu/lbm			Entropy, Btu/lbm·R	
		Sat. Liquid, v_f	Sat. Vapor, v_g	Sat. Liquid, u_f	Sat. Vapor, u_g	Sat. Liquid, h_f	Evap., h_{fg}	Sat. Vapor, h_g	Sat. Liquid, s_f	Sat. Vapor, s_g
−40	7.4272	0.011299	5.7839	63.718	152.95	63.733	97.167	160.90	0.19016	0.42169
−30	9.8624	0.011429	4.4330	66.725	154.32	66.746	95.664	162.41	0.19724	0.41989
−20	12.898	0.011565	3.4449	69.756	155.69	69.784	94.136	163.92	0.20421	0.41831
−15	14.671	0.011635	3.0514	71.281	156.38	71.313	93.357	164.67	0.20766	0.41760
−10	16.632	0.011706	2.7109	72.812	157.06	72.848	92.562	165.41	0.21109	0.41693
−5	18.794	0.011779	2.4154	74.350	157.75	74.391	91.759	166.15	0.21449	0.41631
0	21.171	0.011853	2.1579	75.894	158.43	75.940	90.950	166.89	0.21786	0.41572
5	23.777	0.011929	1.9330	77.445	159.11	77.497	90.123	167.62	0.22122	0.41518
10	26.628	0.012007	1.7357	79.002	159.79	79.062	89.288	168.35	0.22456	0.41467
15	29.739	0.012086	1.5623	80.567	160.47	80.634	88.436	169.07	0.22787	0.41419
20	33.124	0.012168	1.4094	82.140	161.14	82.214	87.576	169.79	0.23117	0.41374
25	36.800	0.012251	1.2742	83.720	161.82	83.803	86.697	170.50	0.23445	0.41332
30	40.784	0.012337	1.1543	85.307	162.49	85.401	85.799	171.20	0.23771	0.41293
35	45.092	0.012425	1.0478	86.903	163.15	87.007	84.893	171.90	0.24095	0.41257
40	49.741	0.012515	0.95280	88.507	163.81	88.623	83.967	172.59	0.24418	0.41222
45	54.749	0.012608	0.86796	90.120	164.47	90.248	83.022	173.27	0.24739	0.41190
50	60.134	0.012703	0.79198	91.742	165.12	91.883	82.057	173.94	0.25059	0.41159
55	65.913	0.012802	0.72380	93.372	165.77	93.529	81.071	174.60	0.25378	0.41131
60	72.105	0.012903	0.66246	95.013	166.41	95.185	80.075	175.26	0.25695	0.41103
65	78.729	0.013008	0.60718	96.663	167.05	96.853	79.047	175.90	0.26011	0.41077
70	85.805	0.013116	0.55724	98.324	167.67	98.532	77.998	176.53	0.26327	0.41052
75	93.351	0.013229	0.51204	99.995	168.30	100.22	76.930	177.15	0.26641	0.41028
80	101.39	0.013345	0.47104	101.68	168.91	101.93	75.820	177.75	0.26955	0.41005
85	109.93	0.013465	0.43379	103.37	169.51	103.65	74.690	178.34	0.27268	0.40982
90	119.01	0.013590	0.39988	105.08	170.11	105.38	73.540	178.92	0.27580	0.40959
95	128.65	0.013720	0.36896	106.80	170.69	107.13	72.350	179.48	0.27892	0.40937
100	138.85	0.013856	0.34070	108.53	171.26	108.89	71.130	180.02	0.28204	0.40914
105	149.65	0.013998	0.31483	110.28	171.82	110.67	69.880	180.55	0.28515	0.40891
110	161.07	0.014146	0.29111	112.04	172.37	112.46	68.590	181.05	0.28827	0.40867
115	173.14	0.014301	0.26933	113.82	172.90	114.28	67.250	181.53	0.29139	0.40842
120	185.86	0.014464	0.24928	115.62	173.41	116.12	65.870	181.99	0.29451	0.40815
140	243.92	0.015214	0.18332	123.00	175.26	123.69	59.850	183.54	0.30708	0.40689
160	314.73	0.016191	0.13428	130.79	176.64	131.74	52.720	184.46	0.31995	0.40503
180	400.34	0.017588	0.096375	139.24	177.19	140.54	43.790	184.33	0.33350	0.40196
200	503.59	0.020096	0.064663	149.07	175.80	150.95	30.880	181.83	0.34896	0.39577
210	563.35	0.023150	0.047695	155.85	172.50	158.26	19.220	177.48	0.35960	0.38830

Table AE.2 Saturation Properties of Refrigerant R-134a: Pressure Increments

Press., psia	Temp., T_{sat} °F	Specific Volume, ft³/lbm		Internal Energy, Btu/lbm		Enthalpy, Btu/lbm			Entropy, Btu/lbm·R	
		Sat. Liquid, v_f	Sat. Vapor, v_g	Sat. Liquid, u_f	Sat. Vapor, u_g	Sat. Liquid, h_f	Evap., h_{fg}	Sat. Vapor, h_g	Sat. Liquid, s_f	Sat. Vapor, s_g
5	−53.066	0.011135	8.3793	59.821	151.16	59.832	99.088	158.92	0.18072	0.42441
10	−29.497	0.011436	4.3756	66.877	154.39	66.898	95.592	162.49	0.19759	0.41980
15	−14.125	0.011647	2.9882	71.548	156.50	71.581	93.219	164.80	0.20826	0.41748
20	−2.4046	0.011817	2.2774	75.150	158.10	75.194	91.346	166.54	0.21624	0.41600
25	7.1984	0.011963	1.8431	78.129	159.41	78.184	89.756	167.94	0.22269	0.41495
30	15.401	0.012093	1.5493	80.693	160.52	80.760	88.370	169.13	0.22814	0.41415
40	29.047	0.012320	1.1761	85.004	162.36	85.095	85.975	171.07	0.23709	0.41301
50	40.267	0.012520	0.94802	88.593	163.85	88.709	83.921	172.63	0.24435	0.41220
60	49.880	0.012701	0.79371	91.703	165.11	91.844	82.086	173.93	0.25051	0.41160
70	58.338	0.012869	0.68214	94.467	166.20	94.633	80.407	175.04	0.25590	0.41112
80	65.922	0.013028	0.59758	96.969	167.16	97.162	78.858	176.02	0.26070	0.41073
90	72.819	0.013179	0.53121	99.265	168.03	99.484	77.396	176.88	0.26504	0.41039
100	79.159	0.013325	0.47767	101.39	168.81	101.64	76.010	177.65	0.26902	0.41009
120	90.526	0.013604	0.39650	105.26	170.17	105.56	73.420	178.98	0.27613	0.40957
140	100.55	0.013871	0.33777	108.72	171.32	109.08	71.000	180.08	0.28238	0.40911
160	109.54	0.014132	0.29321	111.88	172.32	112.30	68.700	181.00	0.28798	0.40869
180	117.73	0.014389	0.25818	114.80	173.18	115.28	66.510	181.79	0.29309	0.40827
200	125.26	0.014645	0.22988	117.53	173.93	118.07	64.380	182.45	0.29780	0.40786
220	132.25	0.014902	0.20649	120.10	174.59	120.70	62.300	183.00	0.30218	0.40743
240	138.77	0.015163	0.18682	122.54	175.16	123.21	60.250	183.46	0.30630	0.40698
260	144.89	0.015428	0.17001	124.86	175.65	125.61	58.230	183.84	0.31019	0.40651
280	150.67	0.015699	0.15546	127.10	176.07	127.91	56.220	184.13	0.31389	0.40600
300	156.15	0.015978	0.14271	129.25	176.42	130.14	54.210	184.35	0.31743	0.40546
350	168.70	0.016727	0.11671	134.36	177.01	135.45	49.130	184.58	0.32572	0.40390
400	179.93	0.017582	0.096497	139.20	177.19	140.51	43.820	184.33	0.33345	0.40197
450	190.09	0.018614	0.079969	143.92	176.88	145.47	38.070	183.54	0.34089	0.39948
500	199.37	0.019977	0.065644	148.72	175.91	150.57	31.410	181.98	0.34840	0.39607

Table AE.3   Superheated Vapor

Table AE.3A $P = 10$ psia ($T_{sat} = -29.497°F$)

T, °F	v, ft³/lbm	u, Btu/lbm	h, Btu/lbm	s, Btu/lbm·R
Sat.	4.3756	154.39	162.49	0.4198
−20	4.4856	155.93	164.23	0.42382
0	4.7136	159.21	167.94	0.43206
20	4.938	162.57	171.71	0.44009
40	5.1601	166	175.56	0.44795
60	5.3802	169.52	179.49	0.45565
80	5.599	173.13	183.5	0.46323
100	5.8166	176.82	187.59	0.47068
120	6.0332	180.6	191.78	0.47802
140	6.249	184.47	196.04	0.48526
160	6.4643	188.43	200.4	0.4924
180	6.6789	192.47	204.84	0.49945
200	6.8931	196.6	209.36	0.50641
220	7.1069	200.81	213.97	0.5133

Table AE.3B $P = 15$ psia ($T_{sat} = -14.125°F$)

T, °F	v, ft³/lbm	u, Btu/lbm	h, Btu/lbm	s, Btu/lbm·R
Sat.	2.9882	156.5	164.8	0.41748
0	3.1001	158.87	167.49	0.42342
20	3.2552	162.28	171.32	0.43159
40	3.4074	165.76	175.22	0.43955
60	3.5577	169.3	179.19	0.44734
80	3.7065	172.94	183.23	0.45497
100	3.854	176.65	187.35	0.46247
120	4.0006	180.45	191.56	0.46985
140	4.1464	184.33	195.85	0.47712
160	4.2916	188.3	200.22	0.48429
180	4.4361	192.35	204.67	0.49137
200	4.5802	196.48	209.21	0.49835
220	4.7239	200.71	213.83	0.50525

Table AE.3C $P = 20$ psia ($T_{sat} = -2.4046°F$)

T, °F	v, ft³/lbm	u, Btu/lbm	h, Btu/lbm	s, Btu/lbm·R
Sat.	2.2774	158.1	166.54	0.416
0	2.2922	158.52	167.01	0.41702
20	2.413	161.99	170.92	0.42536
40	2.5306	165.5	174.87	0.43343
60	2.6461	169.08	178.88	0.4413
80	2.76	172.74	182.96	0.449
100	2.8726	176.47	187.11	0.45655
120	2.9842	180.29	191.34	0.46397
140	3.095	184.18	195.65	0.47128
160	3.2051	188.16	200.03	0.47848
180	3.3147	192.23	204.5	0.48557
200	3.4237	196.37	209.05	0.49258
220	3.5324	200.6	213.68	0.49949

Table AE.3D $P = 30$ psia ($T_{sat} = 15.401°F$)

T, °F	v, ft³/lbm	u, Btu/lbm	h, Btu/lbm	s, Btu/lbm·R
Sat.	1.5493	160.52	169.13	0.41415
20	1.5691	161.35	170.07	0.41612
40	1.6528	164.97	174.15	0.42446
60	1.7338	168.62	178.26	0.43251
80	1.813	172.33	182.41	0.44035
100	1.8908	176.11	186.62	0.44801
120	1.9675	179.96	190.89	0.45551
140	2.0434	183.89	195.24	0.46289
160	2.1185	187.9	199.67	0.47014
180	2.1931	191.98	204.16	0.47729
200	2.2672	196.15	208.74	0.48434
220	2.3408	200.39	213.4	0.49129

Table AE.3E $P = 40$ psia ($T_{sat} = 29.047°F$)

T, °F	v, ft³/lbm	u, Btu/lbm	h, Btu/lbm	s, Btu/lbm·R
Sat.	1.1761	162.36	171.07	0.41301
40	1.2126	164.41	173.39	0.4177
60	1.2768	168.15	177.6	0.42597
80	1.3389	171.92	181.83	0.43396
100	1.3995	175.74	186.11	0.44173
120	1.4589	179.63	190.44	0.44933
140	1.5173	183.59	194.83	0.45678
160	1.5751	187.63	199.29	0.4641
180	1.6322	191.73	203.82	0.4713
200	1.6888	195.92	208.43	0.47839
220	1.7449	200.18	213.11	0.48537
240	1.8007	204.53	217.86	0.49227
260	1.8562	208.95	222.7	0.49908
280	1.9114	213.45	227.61	0.50581

Table AE.3F $P = 50$ psia ($T_{sat} = 40.267°F$)

T, °F	v, ft³/lbm	u, Btu/lbm	h, Btu/lbm	s, Btu/lbm·R
Sat.	0.94802	163.85	172.63	0.4122
60	1.0019	167.64	176.92	0.42063
80	1.054	171.49	181.24	0.42879
100	1.1043	175.36	185.59	0.4367
120	1.1534	179.29	189.97	0.4444
140	1.2015	183.29	194.41	0.45192
160	1.2488	187.35	198.91	0.45931
180	1.2955	191.48	203.48	0.46656
200	1.3416	195.69	208.11	0.47369
220	1.3873	199.97	212.82	0.48071
240	1.4326	204.33	217.59	0.48764
260	1.4776	208.76	222.44	0.49448
280	1.5223	213.28	227.37	0.50123

Table AE.3G $P = 60$ psia ($T_{sat} = 49.880°F$)

T, °F	v, ft³/lbm	u, Btu/lbm	h, Btu/lbm	s, Btu/lbm·R
Sat.	0.79371	165.11	173.93	0.4116
60	0.8179	167.11	176.2	0.41601
80	0.86356	171.03	180.63	0.42438
100	0.90725	174.97	185.05	0.43243
120	0.94955	178.95	189.5	0.44024
140	0.99079	182.98	193.99	0.44785
160	1.0312	187.07	198.53	0.4553
180	1.0709	191.23	203.13	0.4626
200	1.1101	195.46	207.79	0.46978
220	1.1489	199.76	212.52	0.47684
240	1.1872	204.13	217.32	0.4838
260	1.2252	208.58	222.19	0.49067
280	1.2629	213.1	227.14	0.49745
300	1.3004	217.71	232.15	0.50414

Table AE.3H $P = 70$ psia ($T_{sat} = 58.338°F$)

T, °F	v, ft³/lbm	u, Btu/lbm	h, Btu/lbm	s, Btu/lbm·R
Sat.	0.68214	166.2	175.04	0.41112
60	0.68572	166.54	175.43	0.41186
80	0.7271	170.56	179.99	0.42048
100	0.76618	174.57	184.5	0.42869
120	0.80372	178.59	189.01	0.43661
140	0.84013	182.66	193.55	0.44431
160	0.87565	186.78	198.13	0.45183
180	0.91047	190.97	202.77	0.45919
200	0.94472	195.22	207.46	0.46642
220	0.97849	199.54	212.22	0.47352
240	1.0119	203.93	217.05	0.48051
260	1.0449	208.39	221.94	0.48741
280	1.0776	212.93	226.9	0.49421
300	1.1101	217.54	231.93	0.50092

Table AE.3I $P = 80$ psia ($T_{sat} = 65.922°F$)

T, °F	v, ft³/lbm	u, Btu/lbm	h, Btu/lbm	s, Btu/lbm·R
Sat.	0.59758	167.16	176.02	0.41073
80	0.6243	170.07	179.32	0.41693
100	0.66009	174.15	183.93	0.42531
120	0.69416	178.23	188.51	0.43336
140	0.72699	182.34	193.11	0.44116
160	0.75888	186.49	197.74	0.44875
180	0.79004	190.71	202.41	0.45617
200	0.8206	194.98	207.14	0.46345
220	0.85066	199.32	211.92	0.47059
240	0.88031	203.73	216.77	0.47762
260	0.90962	208.21	221.68	0.48454
280	0.93862	212.76	226.66	0.49137
300	0.96737	217.38	231.71	0.49811
320	0.9959	222.08	236.83	0.50476

Table AE.3J $P = 90$ psia ($T_{sat} = 72.819°F$)

T, °F	v, ft³/lbm	u, Btu/lbm	h, Btu/lbm	s, Btu/lbm·R
Sat.	0.53121	168.03	176.88	0.41039
80	0.54389	169.55	178.61	0.41362
100	0.57729	173.71	183.33	0.42221
120	0.60874	177.85	188	0.4304
140	0.63885	182.01	192.65	0.4383
160	0.66796	186.2	197.33	0.44597
180	0.69629	190.44	202.04	0.45345
200	0.724	194.74	206.8	0.46078
220	0.7512	199.1	211.62	0.46797
240	0.77797	203.52	216.49	0.47503
260	0.80438	208.02	221.42	0.48198
280	0.83049	212.58	226.42	0.48884
300	0.85634	217.22	231.49	0.49559
320	0.88196	221.92	236.62	0.50227

Table AE.3K $P = 100$ psia ($T_{sat} = 79.159°F$)

T, °F	v, ft³/lbm	u, Btu/lbm	h, Btu/lbm	s, Btu/lbm·R
Sat.	0.47767	168.81	177.65	0.41009
80	0.47907	168.99	177.86	0.41047
100	0.51077	173.26	182.72	0.41931
120	0.54023	177.47	187.47	0.42766
140	0.56822	181.67	192.19	0.43566
160	0.59513	185.9	196.92	0.44342
180	0.62123	190.17	201.67	0.45097
200	0.64667	194.49	206.47	0.45835
220	0.67158	198.87	211.31	0.46558
240	0.69606	203.32	216.2	0.47268
260	0.72017	207.83	221.16	0.47966
280	0.74397	212.4	226.18	0.48654
300	0.7675	217.05	231.26	0.49332
320	0.7908	221.77	236.41	0.50001

Table AE.3L $P = 120$ psia ($T_{sat} = 90.526°F$)

T, °F	v, ft³/lbm	u, Btu/lbm	h, Btu/lbm	s, Btu/lbm·R
Sat.	0.3965	170.17	178.98	0.40957
100	0.41013	172.28	181.4	0.41393
120	0.43693	176.65	186.36	0.42264
140	0.46191	180.97	191.23	0.4309
160	0.48564	185.28	196.07	0.43884
180	0.50845	189.61	200.91	0.44653
200	0.53054	193.99	205.78	0.45402
220	0.55207	198.41	210.68	0.46134
240	0.57312	202.9	215.63	0.46852
260	0.59379	207.44	220.63	0.47557
280	0.61414	212.04	225.69	0.4825
300	0.6342	216.72	230.81	0.48933
320	0.65403	221.46	235.99	0.49606

Table AE.3M $P = 140$ psia ($T_{sat} = 100.55°F$)

T, °F	v, ft³/lbm	u, Btu/lbm	h, Btu/lbm	s, Btu/lbm·R
Sat.	0.33777	171.32	180.08	0.40911
120	0.36243	175.77	185.17	0.41805
140	0.38551	180.22	190.22	0.4266
160	0.40711	184.63	195.18	0.43475
180	0.42767	189.04	200.12	0.4426
200	0.44743	193.47	205.07	0.45022
220	0.46658	197.95	210.04	0.45764
240	0.48522	202.47	215.05	0.4649
260	0.50346	207.04	220.1	0.47201
280	0.52135	211.68	225.2	0.479
300	0.53895	216.38	230.35	0.48588
320	0.55631	221.14	235.56	0.49265
340	0.57345	225.97	240.84	0.49933

Table AE.3N $P = 160$ psia ($T_{sat} = 109.54°F$)

T, °F	v, ft³/lbm	u, Btu/lbm	h, Btu/lbm	s, Btu/lbm·R
Sat.	0.29321	172.32	181	0.40869
120	0.30578	174.82	183.87	0.41368
140	0.32774	179.43	189.14	0.42262
160	0.34791	183.95	194.26	0.43101
180	0.36687	188.44	199.31	0.43904
200	0.38494	192.94	204.34	0.44679
220	0.40235	197.47	209.39	0.45431
240	0.41921	202.03	214.45	0.46166
260	0.43564	206.64	219.55	0.46884
280	0.45171	211.31	224.69	0.47589
300	0.46748	216.03	229.88	0.48282
320	0.48299	220.82	235.13	0.48964
340	0.49829	225.67	240.44	0.49635

Table AE.3O $P = 180$ psia ($T_{sat} = 117.73°F$)

T, °F	v, ft³/lbm	u, Btu/lbm	h, Btu/lbm	s, Btu/lbm·R
Sat.	0.25818	173.18	181.79	0.40827
120	0.26083	173.75	182.44	0.40941
140	0.28231	178.58	187.99	0.41882
160	0.30154	183.23	193.28	0.4275
180	0.31936	187.82	198.46	0.43573
200	0.33619	192.39	203.6	0.44363
220	0.35228	196.97	208.71	0.45127
240	0.36779	201.58	213.84	0.45871
260	0.38284	206.23	218.99	0.46597
280	0.39751	210.93	224.18	0.47308
300	0.41186	215.69	229.41	0.48006
320	0.42595	220.5	234.69	0.48692
340	0.43981	225.37	240.03	0.49368

Table AE.3P $P = 200$ psia ($T_{sat} = 125.26°F$)

T, °F	v, ft³/lbm	u, Btu/lbm	h, Btu/lbm	s, Btu/lbm·R
Sat.	0.22988	173.93	182.45	0.40786
140	0.24541	177.66	186.75	0.41512
160	0.26412	182.48	192.26	0.42416
180	0.28115	187.17	197.58	0.43262
200	0.29705	191.82	202.82	0.44068
220	0.31213	196.46	208.02	0.44845
240	0.32658	201.12	213.22	0.45599
260	0.34055	205.82	218.43	0.46333
280	0.35411	210.55	223.66	0.4705
300	0.36734	215.33	228.94	0.47754
320	0.38029	220.17	234.25	0.48444
340	0.39301	225.06	239.62	0.49124

Table AE.3Q $P = 300$ psia ($T_{sat} = 156.15°F$)

T, °F	v, ft³/lbm	u, Btu/lbm	h, Btu/lbm	s, Btu/lbm·R
Sat.	0.14271	176.42	184.35	0.40546
160	0.14656	177.63	185.77	0.40776
180	0.16356	183.34	192.42	0.41833
200	0.17777	188.6	198.48	0.42765
220	0.19044	193.67	204.25	0.43627
240	0.20211	198.65	209.88	0.44443
260	0.21306	203.59	215.43	0.45226
280	0.22347	208.53	220.95	0.45982
300	0.23346	213.48	226.45	0.46716
320	0.2431	218.46	231.97	0.47433
340	0.25247	223.48	237.51	0.48134

Table AE.3R $P = 400$ psia ($T_{sat} = 179.93°F$)

T, °F	v, ft³/lbm	u, Btu/lbm	h, Btu/lbm	s, Btu/lbm·R
Sat.	0.096497	177.19	184.33	0.40197
180	0.096581	177.22	184.37	0.40203
200	0.1144	184.33	192.81	0.41502
220	0.12746	190.26	199.7	0.42532
240	0.13853	195.77	206.03	0.4345
260	0.14844	201.09	212.08	0.44303
280	0.15756	206.31	217.98	0.45111
300	0.16611	211.48	223.78	0.45886
320	0.17423	216.64	229.54	0.46634
340	0.18201	221.81	235.29	0.47361

Appendix BE
Thermodynamic Properties of Ideal Gases and Carbon

Table BE.1   CO (Molecular Weight = 28.010, Enthalpy of Formation at 298 K = –47,532.8 Btu/lbmol)

T (R)	$\bar{c}_p$ (Btu/lbmol·R)	$\bar{h}°(T) - \bar{h}_f°(298)$ (Btu/lbmol)	$\bar{s}°(T)$ (Btu/lbmol·R)
300	6.813	−1628.916	43.184
320	6.826	−1492.527	43.624
340	6.839	−1355.872	44.038
360	6.852	−1218.962	44.429
380	6.864	−1081.810	44.800
400	6.875	−944.425	45.153
420	6.886	−806.817	45.488
440	6.896	−668.994	45.809
460	6.907	−530.964	46.116
480	6.917	−392.733	46.410
500	6.926	−254.305	46.692
520	6.936	−115.684	46.964
536.67	6.944	0	47.183
540	6.945	23.125	47.226
560	6.954	162.120	47.479
580	6.964	301.301	47.723
600	6.973	440.667	47.959
620	6.982	580.217	48.188
640	6.991	719.952	48.410
660	7.001	859.874	48.625
680	7.010	999.986	48.834
700	7.020	1140.289	49.038
720	7.030	1280.789	49.236
740	7.040	1421.487	49.428
760	7.050	1562.390	49.616
780	7.061	1703.503	49.799
800	7.072	1844.830	49.978
820	7.083	1986.378	50.153
840	7.095	2128.153	50.324
860	7.106	2270.162	50.491
880	7.119	2412.411	50.655
900	7.131	2554.909	50.815
920	7.144	2697.662	50.972
940	7.158	2840.679	51.125
960	7.171	2983.967	51.276
980	7.186	3127.534	51.424
1000	7.200	3271.389	51.569
1020	7.215	3415.539	51.712
1040	7.230	3559.993	51.852

T (R)	$\bar{c}_p$ (Btu/lbmol·R)	$\bar{h}°(T) - \bar{h}_f°(298)$ (Btu/lbmol)	$\bar{s}°(T)$ (Btu/lbmol·R)
1060	7.246	3704.759	51.990
1080	7.262	3849.845	52.126
1100	7.279	3995.260	52.259
1120	7.296	4141.010	52.391
1140	7.314	4287.105	52.520
1160	7.331	4433.552	52.647
1180	7.349	4580.358	52.773
1200	7.368	4727.530	52.896
1220	7.387	4875.075	53.018
1240	7.406	5023.000	53.139
1260	7.425	5171.310	53.257
1280	7.445	5320.013	53.374
1300	7.465	5469.112	53.490
1320	7.485	5618.612	53.604
1340	7.506	5768.519	53.717
1360	7.526	5918.835	53.828
1380	7.547	6069.564	53.938
1400	7.568	6220.709	54.047
1420	7.589	6372.271	54.154
1440	7.610	6524.252	54.261
1460	7.631	6676.652	54.366
1480	7.651	6829.470	54.470
1500	7.672	6982.706	54.573
1520	7.693	7136.358	54.674
1540	7.714	7290.423	54.775
1560	7.734	7444.898	54.875
1580	7.754	7599.776	54.973
1600	7.774	7755.054	55.071
1620	7.793	7910.725	55.168
1640	7.812	8066.780	55.263
1660	7.831	8223.211	55.358
1680	7.849	8380.008	55.452
1700	7.866	8537.160	55.545
1720	7.883	8694.654	55.637
1740	7.899	8852.478	55.729
1760	7.915	9010.616	55.819
1780	7.929	9169.053	55.908
1800	7.943	9327.771	55.997
1820	7.956	9486.755	56.085
1840	7.969	9645.997	56.172
1860	7.981	9805.496	56.258
1880	7.994	9965.248	56.344
1900	8.006	10125.251	56.428
1940	8.031	10445.999	56.595
1980	8.055	10767.715	56.759
2020	8.078	11090.380	56.921
2060	8.101	11413.971	57.079
2100	8.124	11738.467	57.235
2140	8.145	12063.846	57.389
2180	8.167	12390.090	57.540
2220	8.188	12717.177	57.689
2260	8.208	13045.088	57.835
2300	8.228	13373.803	57.979

(*continued*)

Table BE.1 (continued)

T (R)	$\bar{c}_p$ (Btu/lbmol·R)	$\bar{h}°(T) - \bar{h}_f°(298)$ (Btu/lbmol)	$\bar{s}°(T)$ (Btu/lbmol·R)
2340	8.247	13703.304	58.121
2380	8.266	14033.572	58.261
2420	8.285	14364.588	58.399
2460	8.303	14696.334	58.535
2500	8.320	15028.793	58.669
2540	8.337	15361.948	58.801
2580	8.354	15695.781	58.932
2620	8.371	16030.276	59.060
2660	8.386	16365.417	59.187
2700	8.402	16701.186	59.313
2740	8.417	17037.570	59.436
2780	8.432	17374.552	59.558
2820	8.446	17712.117	59.679
2860	8.460	18050.251	59.798
2900	8.474	18388.939	59.916
2940	8.487	18728.167	60.032
2980	8.500	19067.921	60.147
3020	8.513	19408.187	60.260
3060	8.525	19748.953	60.372
3100	8.537	20090.205	60.483
3140	8.549	20431.931	60.592
3180	8.560	20774.118	60.701
3220	8.571	21116.755	60.808
3260	8.582	21459.828	60.914
3300	8.593	21803.328	61.018
3340	8.603	22147.241	61.122
3380	8.613	22491.559	61.224
3420	8.623	22836.268	61.326
3460	8.632	23181.360	61.426
3500	8.641	23526.824	61.525
3540	8.650	23872.650	61.624
3580	8.659	24218.828	61.721
3620	8.667	24565.349	61.817
3660	8.676	24912.203	61.912
3700	8.684	25259.382	62.007
3740	8.691	25606.877	62.100
3780	8.699	25954.678	62.193
3820	8.706	26302.779	62.284
3860	8.713	26651.170	62.375
3900	8.720	26999.845	62.465
3940	8.727	27348.794	62.554
3980	8.734	27698.012	62.642
4020	8.740	28047.491	62.730
4060	8.746	28397.223	62.816
4100	8.753	28747.202	62.902
4140	8.759	29097.422	62.987
4180	8.764	29447.876	63.071
4220	8.770	29798.558	63.155
4260	8.775	30149.463	63.237
4300	8.781	30500.583	63.319
4340	8.786	30851.914	63.401
4380	8.791	31203.451	63.481
4420	8.796	31555.188	63.561
4460	8.801	31907.120	63.641

T (R)	$\bar{c}_p$ (Btu/lbmol·R)	$\bar{h}°(T) - \bar{h}_f°(298)$ (Btu/lbmol)	$\bar{s}°(T)$ (Btu/lbmol·R)
4500	8.805	32259.242	63.719
4540	8.810	32611.550	63.797
4580	8.815	32964.040	63.874
4620	8.819	33316.706	63.951
4660	8.823	33669.545	64.027
4700	8.827	34022.552	64.103
4740	8.831	34375.725	64.177
4780	8.835	34729.058	64.252
4820	8.839	35082.549	64.325
4860	8.843	35436.195	64.398
4900	8.847	35789.992	64.471
5000	8.856	36675.124	64.650
5100	8.864	37561.137	64.825
5200	8.873	38447.991	64.997
5300	8.881	39335.652	65.166
5400	8.888	40224.088	65.332
5500	8.896	41113.275	65.496
5600	8.903	42003.191	65.656
5700	8.910	42893.815	65.814
5800	8.917	43785.135	65.969
5900	8.923	44677.136	66.121
6000	8.930	45569.809	66.271
6100	8.937	46463.147	66.419
6200	8.943	47357.144	66.564
6300	8.950	48251.796	66.707
6400	8.956	49147.101	66.848
6500	8.963	50043.058	66.987
6600	8.969	50939.665	67.124
6700	8.976	51836.923	67.259
6800	8.982	52734.831	67.392
6900	8.989	53633.389	67.523
7000	8.995	54532.596	67.653
7100	9.002	55432.450	67.780
7200	9.008	56332.948	67.906
7300	9.015	57234.086	68.031
7400	9.021	58135.858	68.153
7500	9.027	59038.256	68.274
7600	9.033	59941.268	68.394
7700	9.039	60844.881	68.512
7800	9.045	61749.080	68.629
7900	9.050	62653.843	68.744
8000	9.056	63559.149	68.858
8100	9.061	64464.969	68.970
8200	9.065	65371.271	69.082
8300	9.070	66278.020	69.192
8400	9.073	67185.174	69.300
8500	9.077	68092.687	69.408
8600	9.080	69000.506	69.514
8700	9.082	69908.574	69.619
8800	9.083	70816.827	69.723
8900	9.084	71725.194	69.825
9000	9.084	72633.598	69.927

Table BE.2 CO_2 (Molecular Weight = 44.011, Enthalpy of Formation at 298 K = −169,193 Btu/lbmol)

T (R)	$\bar{c}_p$ (Btu/lbmol·R)	$\bar{h}°(T) - \bar{h}_f°(298)$ (Btu/lbmol)	$\bar{s}°(T)$ (Btu/lbmol·R)
300	7.289	−1922.431	46.370
320	7.441	−1775.125	46.846
340	7.590	−1624.807	47.301
360	7.736	−1471.546	47.739
380	7.878	−1315.411	48.161
400	8.016	−1156.466	48.569
420	8.152	−994.776	48.963
440	8.285	−830.404	49.346
460	8.414	−663.410	49.717
480	8.541	−493.853	50.077
500	8.665	−321.793	50.429
520	8.786	−147.284	50.771
536.67	8.885	0	51.050
540	8.904	29.618	51.105
560	9.020	208.860	51.431
580	9.133	390.389	51.749
600	9.244	574.156	52.061
620	9.352	760.111	52.365
640	9.458	948.206	52.664
660	9.561	1138.395	52.957
680	9.662	1330.631	53.244
700	9.761	1524.872	53.525
720	9.858	1721.073	53.801
740	9.953	1919.194	54.073
760	10.046	2119.193	54.340
780	10.137	2321.031	54.602
800	10.226	2524.670	54.859
820	10.314	2730.072	55.113
840	10.399	2937.201	55.363
860	10.483	3146.020	55.608
880	10.565	3356.497	55.850
900	10.645	3568.597	56.089
920	10.724	3782.288	56.323
940	10.801	3997.538	56.555
960	10.877	4214.315	56.783
980	10.951	4432.590	57.008
1000	11.023	4652.334	57.230
1020	11.095	4873.517	57.449
1040	11.165	5096.112	57.665
1060	11.233	5320.091	57.878
1080	11.300	5545.429	58.089
1100	11.366	5772.098	58.297
1120	11.431	6000.073	58.502
1140	11.495	6229.330	58.705
1160	11.557	6459.844	58.906
1180	11.618	6691.591	59.104
1200	11.678	6924.548	59.300
1220	11.737	7158.692	59.493
1240	11.794	7394.001	59.684
1260	11.851	7630.451	59.874

T (R)	$\bar{c}_p$ (Btu/lbmol·R)	$\bar{h}°(T) - \bar{h}_f°(298)$ (Btu/lbmol)	$\bar{s}°(T)$ (Btu/lbmol·R)
1280	11.906	7868.022	60.061
1300	11.961	8106.691	60.246
1320	12.014	8346.438	60.429
1340	12.066	8587.241	60.610
1360	12.118	8829.079	60.789
1380	12.168	9071.933	60.966
1400	12.217	9315.781	61.142
1420	12.265	9560.602	61.315
1440	12.312	9806.378	61.487
1460	12.359	10053.087	61.657
1480	12.404	10300.709	61.826
1500	12.448	10549.224	61.992
1520	12.491	10798.613	62.158
1540	12.533	11048.854	62.321
1560	12.574	11299.927	62.483
1580	12.614	11551.812	62.644
1600	12.653	11804.487	62.803
1620	12.691	12057.934	62.960
1640	12.728	12312.129	63.116
1660	12.764	12567.052	63.270
1680	12.799	12822.682	63.423
1700	12.833	13078.996	63.575
1720	12.865	13335.972	63.725
1740	12.896	13593.587	63.874
1760	12.927	13851.820	64.022
1780	12.956	14110.645	64.168
1800	12.984	14370.040	64.313
1820	13.011	14629.950	64.457
1840	13.037	14890.431	64.599
1860	13.064	15151.444	64.740
1880	13.090	15412.983	64.880
1900	13.116	15675.041	65.019
1940	13.167	16200.691	65.292
1980	13.216	16728.345	65.562
2020	13.264	17257.955	65.826
2060	13.311	17789.472	66.087
2100	13.357	18322.852	66.343
2140	13.402	18858.047	66.596
2180	13.446	19395.014	66.845
2220	13.489	19933.708	67.089
2260	13.530	20474.086	67.331
2300	13.571	21016.105	67.568
2340	13.610	21559.725	67.803
2380	13.649	22104.903	68.034
2420	13.686	22651.602	68.262
2460	13.723	23199.781	68.486
2500	13.758	23749.401	68.708
2540	13.793	24300.427	68.926
2580	13.827	24852.819	69.142
2620	13.860	25406.544	69.355
2660	13.892	25961.564	69.565
2700	13.923	26517.846	69.773

(continued)

Table BE.2   (continued)

T (R)	$\bar{c}_p$ (Btu/lbmol·R)	$\bar{h}°(T) - \bar{h}_f°(298)$ (Btu/lbmol)	$\bar{s}°(T)$ (Btu/lbmol·R)
2740	13.953	27075.356	69.978
2780	13.982	27634.061	70.180
2820	14.011	28193.929	70.380
2860	14.039	28754.927	70.578
2900	14.066	29317.024	70.773
2940	14.092	29880.192	70.966
2980	14.118	30444.399	71.157
3020	14.143	31009.617	71.345
3060	14.167	31575.818	71.531
3100	14.191	32142.975	71.715
3140	14.214	32711.060	71.897
3180	14.236	33280.047	72.078
3220	14.257	33849.911	72.256
3260	14.278	34420.626	72.432
3300	14.299	34992.169	72.606
3340	14.319	35564.515	72.778
3380	14.338	36137.642	72.949
3420	14.356	36711.527	73.118
3460	14.375	37286.148	73.285
3500	14.392	37861.483	73.450
3540	14.409	38437.513	73.614
3580	14.426	39014.217	73.776
3620	14.442	39591.575	73.936
3660	14.458	40169.568	74.095
3700	14.473	40748.178	74.252
3740	14.488	41327.387	74.408
3780	14.502	41907.178	74.562
3820	14.516	42487.533	74.715
3860	14.529	43068.436	74.866
3900	14.543	43649.872	75.016
3940	14.555	44231.825	75.164
3980	14.568	44814.280	75.312
4020	14.580	45397.223	75.457
4060	14.591	45980.639	75.602
4100	14.603	46564.517	75.745
4140	14.614	47148.841	75.887
4180	14.624	47733.601	76.027
4220	14.635	48318.784	76.167
4260	14.645	48904.379	76.305
4300	14.655	49490.373	76.442
4340	14.664	50076.758	76.577
4380	14.674	50663.521	76.712
4420	14.683	51250.654	76.845
4460	14.692	51838.147	76.978
4500	14.700	52425.990	77.109
4540	14.709	53014.175	77.239
4580	14.717	53602.694	77.368
4620	14.725	54191.539	77.496
4660	14.733	54780.701	77.623
4700	14.741	55370.175	77.749
4740	14.748	55959.952	77.874
4780	14.756	56550.027	77.998

T (R)	$\bar{c}_p$ (Btu/lbmol·R)	$\bar{h}°(T) - \bar{h}_f°(298)$ (Btu/lbmol)	$\bar{s}°(T)$ (Btu/lbmol·R)
4820	14.763	57140.392	78.121
4860	14.770	57731.043	78.243
4900	14.777	58321.973	78.364
5000	14.794	59800.488	78.663
5100	14.810	61280.644	78.956
5200	14.825	62762.376	79.244
5300	14.840	64245.633	79.526
5400	14.855	65730.369	79.804
5500	14.869	67216.551	80.076
5600	14.883	68704.151	80.344
5700	14.897	70193.149	80.608
5800	14.911	71683.533	80.867
5900	14.925	73175.295	81.122
6000	14.938	74668.432	81.373
6100	14.952	76162.946	81.620
6200	14.966	77658.842	81.863
6300	14.980	79156.127	82.103
6400	14.994	80654.812	82.339
6500	15.008	82154.907	82.572
6600	15.022	83656.424	82.801
6700	15.037	85159.373	83.027
6800	15.051	86663.765	83.250
6900	15.066	88169.607	83.470
7000	15.080	89676.905	83.686
7100	15.095	91185.660	83.900
7200	15.109	92695.871	84.112
7300	15.124	94207.531	84.320
7400	15.138	95720.626	84.526
7500	15.152	97235.138	84.729
7600	15.166	98751.041	84.930
7700	15.179	100268.300	85.128
7800	15.192	101786.872	85.324
7900	15.205	103306.706	85.518
8000	15.216	104827.738	85.709
8100	15.227	106349.895	85.898
8200	15.237	107873.093	86.085
8300	15.246	109397.233	86.270
8400	15.254	110922.205	86.453
8500	15.260	112447.884	86.633
8600	15.265	113974.130	86.812
8700	15.268	115500.788	86.988
8800	15.270	117027.687	87.163
8900	15.269	118554.639	87.335
9000	15.267	120081.437	87.506

Table BE.3 H$_2$ (Molecular Weight = 2.016, Enthalpy of Formation at 298 K = 0 Btu/lbmol)

T (R)	$\bar{c}_p$ (Btu/lbmol·R)	$\bar{h}°(T) - \bar{h}_f°(298)$ (Btu/lbmol)	$\bar{s}°(T)$ (Btu/lbmol·R)
300	6.779	−1620.222	27.217
320	6.792	−1484.516	27.655
340	6.804	−1348.564	28.067
360	6.815	−1212.375	28.456
380	6.827	−1075.954	28.825
400	6.838	−939.309	29.175
420	6.849	−802.445	29.509
440	6.859	−665.369	29.828
460	6.869	−528.086	30.133
480	6.879	−390.602	30.426
500	6.889	−252.923	30.707
520	6.898	−115.055	30.977
536.67	6.906	0	31.195
540	6.907	22.998	31.238
560	6.916	161.231	31.489
580	6.925	299.638	31.732
600	6.933	438.215	31.967
620	6.941	576.958	32.194
640	6.949	715.861	32.415
660	6.957	854.922	32.629
680	6.965	994.137	32.837
700	6.972	1133.501	33.039
720	6.979	1273.011	33.235
740	6.986	1412.665	33.426
760	6.993	1552.460	33.613
780	7.000	1692.392	33.795
800	7.007	1832.460	33.972
820	7.013	1972.662	34.145
840	7.020	2112.995	34.314
860	7.026	2253.458	34.479
880	7.033	2394.051	34.641
900	7.039	2534.771	34.799
920	7.046	2675.619	34.954
940	7.052	2816.593	35.105
960	7.058	2957.695	35.254
980	7.065	3098.923	35.400
1000	7.071	3240.280	35.542
1020	7.078	3381.765	35.682
1040	7.084	3523.380	35.820
1060	7.091	3665.126	35.955
1080	7.097	3807.006	36.088
1100	7.104	3949.022	36.218
1120	7.111	4091.175	36.346
1140	7.118	4233.471	36.472
1160	7.126	4375.910	36.596
1180	7.133	4518.498	36.718
1200	7.141	4661.239	36.838
1220	7.149	4804.136	36.956
1240	7.157	4947.196	37.072
1260	7.166	5090.422	37.187
1280	7.174	5233.822	37.299

T (R)	$\bar{c}_p$ (Btu/lbmol·R)	$\bar{h}°(T) - \bar{h}_f°(298)$ (Btu/lbmol)	$\bar{s}°(T)$ (Btu/lbmol·R)
1300	7.184	5377.400	37.411
1320	7.193	5521.163	37.521
1340	7.203	5665.119	37.629
1360	7.213	5809.275	37.736
1380	7.224	5953.638	37.841
1400	7.235	6098.217	37.945
1420	7.246	6243.020	38.048
1440	7.258	6388.057	38.149
1460	7.270	6533.337	38.249
1480	7.283	6678.870	38.348
1500	7.297	6824.667	38.446
1520	7.311	6970.738	38.543
1540	7.325	7117.096	38.638
1560	7.340	7263.751	38.733
1580	7.356	7410.718	38.827
1600	7.373	7558.007	38.919
1620	7.390	7705.634	39.011
1640	7.408	7853.612	39.102
1660	7.427	8001.955	39.192
1680	7.446	8150.679	39.281
1700	7.466	8299.798	39.369
1720	7.487	8449.330	39.456
1740	7.509	8599.290	39.543
1760	7.532	8749.696	39.629
1780	7.555	8900.566	39.714
1800	7.580	9051.918	39.799
1820	7.216	9026.766	39.753
1840	7.229	9171.212	39.832
1860	7.241	9315.907	39.910
1880	7.253	9460.849	39.988
1900	7.266	9606.038	40.065
1940	7.290	9897.153	40.216
1980	7.315	10189.247	40.365
2020	7.339	10482.313	40.512
2060	7.363	10776.347	40.656
2100	7.387	11071.341	40.798
2140	7.411	11367.290	40.937
2180	7.434	11664.188	41.075
2220	7.458	11962.029	41.210
2260	7.481	12260.807	41.344
2300	7.504	12560.516	41.475
2340	7.527	12861.151	41.605
2380	7.550	13162.704	41.732
2420	7.573	13465.170	41.858
2460	7.596	13768.544	41.983
2500	7.618	14072.819	42.106
2540	7.640	14377.989	42.227
2580	7.663	14684.048	42.346
2620	7.685	14990.990	42.464
2660	7.706	15298.809	42.581
2700	7.728	15607.500	42.696
2740	7.750	15917.055	42.810

(continued)

Table BE.3 (continued)

T (R)	$\bar{c}_p$ (Btu/lbmol·R)	$\bar{h}°(T) - \bar{h}_f°(298)$ (Btu/lbmol)	$\bar{s}°(T)$ (Btu/lbmol·R)
2780	7.771	16227.470	42.922
2820	7.792	16538.738	43.033
2860	7.813	16850.853	43.143
2900	7.834	17163.809	43.252
2940	7.855	17477.600	43.359
2980	7.876	17792.220	43.466
3020	7.896	18107.664	43.571
3060	7.917	18423.925	43.675
3100	7.937	18740.997	43.778
3140	7.957	19058.874	43.880
3180	7.977	19377.551	43.981
3220	7.997	19697.021	44.080
3260	8.016	20017.278	44.179
3300	8.036	20338.317	44.277
3340	8.055	20660.132	44.374
3380	8.074	20982.717	44.470
3420	8.093	21306.065	44.565
3460	8.112	21630.171	44.659
3500	8.131	21955.030	44.753
3540	8.149	22280.635	44.845
3580	8.168	22606.980	44.937
3620	8.186	22934.060	45.028
3660	8.204	23261.869	45.118
3700	8.222	23590.402	45.207
3740	8.240	23919.651	45.296
3780	8.258	24249.613	45.383
3820	8.276	24580.280	45.470
3860	8.293	24911.647	45.557
3900	8.310	25243.710	45.642
3940	8.327	25576.461	45.727
3980	8.344	25909.895	45.811
4020	8.361	26244.007	45.895
4060	8.378	26578.792	45.978
4100	8.395	26914.243	46.060
4140	8.411	27250.355	46.142
4180	8.427	27587.123	46.223
4220	8.444	27924.541	46.303
4260	8.460	28262.604	46.383
4300	8.476	28601.306	46.462
4340	8.491	28940.643	46.540
4380	8.507	29280.607	46.618
4420	8.523	29621.196	46.696
4460	8.538	29962.402	46.773
4500	8.553	30304.221	46.849
4540	8.568	30646.647	46.925
4580	8.583	30989.676	47.000
4620	8.598	31333.303	47.075
4660	8.613	31677.521	47.149
4700	8.627	32022.326	47.222
4740	8.642	32367.714	47.296
4780	8.656	32713.679	47.368
4820	8.671	33060.215	47.440

T (R)	$\bar{c}_p$ (Btu/lbmol·R)	$\bar{h}°(T) - \bar{h}_f°(298)$ (Btu/lbmol)	$\bar{s}°(T)$ (Btu/lbmol·R)
4860	8.685	33407.319	47.512
4900	8.699	33754.985	47.583
5000	8.733	34626.579	47.760
5100	8.767	35501.583	47.933
5200	8.800	36379.925	48.103
5300	8.832	37261.533	48.271
5400	8.864	38146.338	48.437
5500	8.895	39034.273	48.600
5600	8.925	39925.274	48.760
5700	8.955	40819.277	48.918
5800	8.984	41716.220	49.074
5900	9.013	42616.047	49.228
6000	9.041	43518.699	49.380
6100	9.068	44424.123	49.530
6200	9.095	45332.268	49.677
6300	9.121	46243.084	49.823
6400	9.147	47156.524	49.967
6500	9.173	48072.545	50.109
6600	9.198	48991.106	50.249
6700	9.223	49912.167	50.388
6800	9.248	50835.694	50.524
6900	9.272	51761.654	50.660
7000	9.296	52690.016	50.793
7100	9.319	53620.756	50.925
7200	9.343	54553.848	51.056
7300	9.366	55489.273	51.185
7400	9.389	56427.014	51.312
7500	9.412	57367.057	51.439
7600	9.435	58309.392	51.563
7700	9.458	59254.013	51.687
7800	9.481	60200.917	51.809
7900	9.503	61150.103	51.930
8000	9.526	62101.577	52.050
8100	9.549	63055.346	52.168
8200	9.572	64011.422	52.285
8300	9.596	64969.822	52.402
8400	9.619	65930.565	52.517
8500	9.643	66893.675	52.631
8600	9.667	67859.180	52.744
8700	9.692	68827.113	52.855
8800	9.716	69797.510	52.966
8900	9.742	70770.412	53.076
9000	9.768	71745.865	53.185

Table BE.4   H (Molecular Weight = 1.088, Enthalpy of Formation at 298 K = 93,713.5 Btu/lbmol)

T (R)	$\bar{c}_p$ (Btu/lbmol·R)	$\bar{h}°(T) - \bar{h}_f°(298)$ (Btu/lbmol)	$\bar{s}°(T)$ (Btu/lbmol·R)
300	4.965	−1174.968	24.485
320	4.965	−1075.677	24.806
340	4.965	−976.385	25.107
360	4.965	−877.093	25.390
380	4.965	−777.801	25.659
400	4.965	−678.510	25.913
420	4.965	−579.218	26.156
440	4.965	−479.926	26.387
460	4.965	−380.635	26.607
480	4.965	−281.343	26.819
500	4.965	−182.051	27.021
520	4.965	−82.760	27.216
536.67	4.965	0	27.373
540	4.965	16.532	27.403
560	4.965	115.824	27.584
580	4.965	215.115	27.758
600	4.965	314.407	27.926
620	4.965	413.699	28.089
640	4.965	512.991	28.247
660	4.965	612.282	28.400
680	4.965	711.574	28.548
700	4.965	810.866	28.692
720	4.965	910.157	28.832
740	4.965	1009.449	28.968
760	4.965	1108.741	29.100
780	4.965	1208.032	29.229
800	4.965	1307.324	29.355
820	4.965	1406.616	29.477
840	4.965	1505.908	29.597
860	4.965	1605.199	29.714
880	4.965	1704.491	29.828
900	4.965	1803.783	29.939
920	4.965	1903.074	30.049
940	4.965	2002.366	30.155
960	4.965	2101.658	30.260
980	4.965	2200.949	30.362
1000	4.965	2300.241	30.462
1020	4.965	2399.533	30.561
1040	4.965	2498.824	30.657
1060	4.965	2598.116	30.752
1080	4.965	2697.408	30.845
1100	4.965	2796.700	30.936
1120	4.965	2895.991	31.025
1140	4.965	2995.283	31.113
1160	4.965	3094.575	31.199
1180	4.965	3193.866	31.284
1200	4.965	3293.158	31.368
1220	4.965	3392.450	31.450
1240	4.965	3491.741	31.530
1260	4.965	3591.033	31.610
1280	4.965	3690.325	31.688

T (R)	$\bar{c}_p$ (Btu/lbmol·R)	$\bar{h}°(T) - \bar{h}_f°(298)$ (Btu/lbmol)	$\bar{s}°(T)$ (Btu/lbmol·R)
1300	4.965	3789.617	31.765
1320	4.965	3888.908	31.841
1340	4.965	3988.200	31.915
1360	4.965	4087.492	31.989
1380	4.965	4186.783	32.062
1400	4.965	4286.075	32.133
1420	4.965	4385.367	32.203
1440	4.965	4484.658	32.273
1460	4.965	4583.950	32.341
1480	4.965	4683.242	32.409
1500	4.965	4782.533	32.475
1520	4.965	4881.825	32.541
1540	4.965	4981.117	32.606
1560	4.965	5080.409	32.670
1580	4.965	5179.700	32.733
1600	4.965	5278.992	32.796
1620	4.965	5378.284	32.858
1640	4.965	5477.575	32.918
1660	4.965	5576.867	32.979
1680	4.965	5676.159	33.038
1700	4.965	5775.450	33.097
1720	4.965	5874.742	33.155
1740	4.965	5974.034	33.212
1760	4.965	6073.326	33.269
1780	4.965	6172.617	33.325
1800	4.965	6271.909	33.381
1820	4.965	6371.201	33.435
1840	4.965	6470.492	33.490
1860	4.965	6569.784	33.543
1880	4.965	6669.076	33.597
1900	4.965	6768.367	33.649
1940	4.965	6966.951	33.752
1980	4.965	7165.534	33.854
2020	4.965	7364.118	33.953
2060	4.965	7562.701	34.050
2100	4.965	7761.284	34.146
2140	4.965	7959.868	34.240
2180	4.965	8158.451	34.332
2220	4.965	8357.035	34.422
2260	4.965	8555.618	34.510
2300	4.965	8754.201	34.598
2340	4.965	8952.785	34.683
2380	4.965	9151.368	34.767
2420	4.965	9349.951	34.850
2460	4.965	9548.535	34.931
2500	4.965	9747.118	35.012
2540	4.965	9945.702	35.090
2580	4.965	10144.285	35.168
2620	4.965	10342.868	35.244
2660	4.965	10541.452	35.320
2700	4.965	10740.035	35.394

(*continued*)

Table BE.4 (continued)

T (R)	$\bar{c}_p$ (Btu/lbmol·R)	$\bar{h}°(T) - \bar{h}_f°(298)$ (Btu/lbmol)	$\bar{s}°(T)$ (Btu/lbmol·R)
2740	4.965	10938.619	35.467
2780	4.965	11137.202	35.539
2820	4.965	11335.785	35.609
2860	4.965	11534.369	35.679
2900	4.965	11732.952	35.748
2940	4.965	11931.536	35.816
2980	4.965	12130.119	35.883
3020	4.965	12328.702	35.950
3060	4.965	12527.286	36.015
3100	4.965	12725.869	36.079
3140	4.965	12924.453	36.143
3180	4.965	13123.036	36.206
3220	4.965	13321.619	36.268
3260	4.965	13520.203	36.329
3300	4.965	13718.786	36.390
3340	4.965	13917.369	36.450
3380	4.965	14115.953	36.509
3420	4.965	14314.536	36.567
3460	4.965	14513.120	36.625
3500	4.965	14711.703	36.682
3540	4.965	14910.286	36.738
3580	4.965	15108.870	36.794
3620	4.965	15307.453	36.849
3660	4.965	15506.037	36.904
3700	4.965	15704.620	36.958
3740	4.965	15903.203	37.011
3780	4.965	16101.787	37.064
3820	4.965	16300.370	37.116
3860	4.965	16498.954	37.168
3900	4.965	16697.537	37.219
3940	4.965	16896.120	37.270
3980	4.965	17094.704	37.320
4020	4.965	17293.287	37.370
4060	4.965	17491.871	37.419
4100	4.965	17690.454	37.467
4140	4.965	17889.037	37.516
4180	4.965	18087.621	37.563
4220	4.965	18286.204	37.611
4260	4.965	18484.788	37.658
4300	4.965	18683.371	37.704
4340	4.965	18881.954	37.750
4380	4.965	19080.538	37.795
4420	4.965	19279.121	37.841
4460	4.965	19477.704	37.885
4500	4.965	19676.288	37.930
4540	4.965	19874.871	37.974
4580	4.965	20073.455	38.017
4620	4.965	20272.038	38.060
4660	4.965	20470.621	38.103
4700	4.965	20669.205	38.146
4740	4.965	20867.788	38.188
4780	4.965	21066.372	38.229

T (R)	$\bar{c}_p$ (Btu/lbmol·R)	$\bar{h}°(T) - \bar{h}_f°(298)$ (Btu/lbmol)	$\bar{s}°(T)$ (Btu/lbmol·R)
4820	4.965	21264.955	38.271
4860	4.965	21463.538	38.312
4900	4.965	21662.122	38.352
5000	4.965	22158.580	38.453
5100	4.965	22655.039	38.551
5200	4.965	23151.497	38.647
5300	4.965	23647.956	38.742
5400	4.965	24144.414	38.835
5500	4.965	24640.873	38.926
5600	4.965	25137.331	39.015
5700	4.965	25633.790	39.103
5800	4.965	26130.248	39.190
5900	4.965	26626.707	39.274
6000	4.965	27123.165	39.358
6100	4.965	27619.624	39.440
6200	4.965	28116.082	39.521
6300	4.965	28612.540	39.600
6400	4.965	29108.999	39.678
6500	4.965	29605.457	39.755
6600	4.965	30101.916	39.831
6700	4.965	30598.374	39.906
6800	4.965	31094.833	39.979
6900	4.965	31591.291	40.052
7000	4.965	32087.750	40.123
7100	4.965	32584.208	40.194
7200	4.965	33080.667	40.263
7300	4.965	33577.125	40.332
7400	4.965	34073.584	40.399
7500	4.965	34570.042	40.466
7600	4.965	35066.501	40.531
7700	4.965	35562.959	40.596
7800	4.965	36059.418	40.660
7900	4.965	36555.876	40.724
8000	4.965	37052.335	40.786
8100	4.965	37548.793	40.848
8200	4.965	38045.252	40.909
8300	4.965	38541.710	40.969
8400	4.965	39038.169	41.028
8500	4.965	39534.627	41.087
8600	4.965	40031.086	41.145
8700	4.965	40527.544	41.203
8800	4.965	41024.002	41.259
8900	4.965	41520.461	41.315
9000	4.965	42016.919	41.371

Table BE.5   OH (Molecular Weight = 17.007, Enthalpy of Formation at 298 K = 16769.9 Btu/lbmol)

T (R)	$\bar{c}_p$ (Btu/lbmol·R)	$\bar{h}°(T) - \bar{h}_f°(298)$ (Btu/lbmol)	$\bar{s}°(T)$ (Btu/lbmol·R)
300	7.213	−1699.869	39.674
320	7.208	−1555.663	40.139
340	7.204	−1411.546	40.576
360	7.199	−1267.525	40.987
380	7.194	−1123.603	41.376
400	7.188	−979.785	41.745
420	7.183	−836.074	42.096
440	7.177	−692.474	42.430
460	7.172	−548.986	42.749
480	7.166	−405.613	43.054
500	7.160	−262.356	43.346
520	7.154	−119.217	43.627
536.67	7.149	0	43.853
540	7.148	23.805	43.897
560	7.142	166.709	44.157
580	7.136	309.495	44.407
600	7.131	452.165	44.649
620	7.125	594.719	44.883
640	7.119	737.160	45.109
660	7.114	879.489	45.328
680	7.108	1021.709	45.540
700	7.103	1163.822	45.746
720	7.098	1305.832	45.946
740	7.093	1447.743	46.141
760	7.088	1589.558	46.330
780	7.084	1731.281	46.514
800	7.080	1872.917	46.693
820	7.076	2014.472	46.868
840	7.072	2155.949	47.038
860	7.069	2297.354	47.205
880	7.065	2438.694	47.367
900	7.063	2579.973	47.526
920	7.060	2721.198	47.681
940	7.058	2862.376	47.833
960	7.056	3003.513	47.982
980	7.054	3144.616	48.127
1000	7.053	3285.692	48.270
1020	7.052	3426.747	48.409
1040	7.052	3567.791	48.546
1060	7.052	3708.829	48.680
1080	7.052	3849.869	48.812
1100	7.053	3990.920	48.942
1120	7.054	4131.989	49.069
1140	7.056	4273.085	49.194
1160	7.058	4414.215	49.316
1180	7.060	4555.387	49.437
1200	7.063	4696.610	49.556
1220	7.066	4837.892	49.672
1240	7.069	4979.241	49.787
1260	7.073	5120.667	49.901

T (R)	$\bar{c}_p$ (Btu/lbmol·R)	$\bar{h}°(T) - \bar{h}_f°(298)$ (Btu/lbmol)	$\bar{s}°(T)$ (Btu/lbmol·R)
1280	7.078	5262.176	50.012
1300	7.083	5403.779	50.122
1320	7.088	5545.482	50.230
1340	7.094	5687.295	50.337
1360	7.100	5829.225	50.442
1380	7.106	5971.282	50.545
1400	7.113	6113.473	50.648
1420	7.120	6255.807	50.749
1440	7.128	6398.291	50.848
1460	7.136	6540.934	50.947
1480	7.145	6683.745	51.044
1500	7.154	6826.729	51.140
1520	7.163	6969.896	51.235
1540	7.173	7113.253	51.328
1560	7.183	7256.807	51.421
1580	7.193	7400.566	51.512
1600	7.204	7544.536	51.603
1620	7.215	7688.724	51.693
1640	7.226	7833.138	51.781
1660	7.238	7977.784	51.869
1680	7.250	8122.667	51.956
1700	7.263	8267.794	52.041
1720	7.275	8413.171	52.126
1740	7.288	8558.803	52.211
1760	7.301	8704.696	52.294
1780	7.315	8850.854	52.377
1800	7.328	8997.282	52.458
1820	7.342	9143.986	52.539
1840	7.356	9290.961	52.620
1860	7.369	9438.206	52.699
1880	7.383	9585.721	52.778
1900	7.396	9733.504	52.856
1940	7.422	10029.867	53.011
1980	7.448	10327.283	53.163
2020	7.474	10625.739	53.312
2060	7.500	10925.223	53.459
2100	7.525	11225.722	53.603
2140	7.550	11527.224	53.745
2180	7.575	11829.718	53.885
2220	7.599	12133.190	54.023
2260	7.623	12437.629	54.159
2300	7.647	12743.022	54.293
2340	7.670	13049.360	54.425
2380	7.693	13356.628	54.555
2420	7.716	13664.817	54.684
2460	7.739	13973.914	54.811
2500	7.761	14283.908	54.936
2540	7.783	14594.789	55.059
2580	7.805	14906.544	55.181
2620	7.826	15219.162	55.301
2660	7.847	15532.634	55.420

(*continued*)

Table BE.5 (continued)

T (R)	$\bar{c}_p$ (Btu/lbmol·R)	$\bar{h}^\circ(T) - \bar{h}_f^\circ(298)$ (Btu/lbmol)	$\bar{s}^\circ(T)$ (Btu/lbmol·R)
2700	7.868	15846.947	55.537
2740	7.889	16162.091	55.653
2780	7.909	16478.055	55.767
2820	7.929	16794.830	55.880
2860	7.949	17112.403	55.992
2900	7.969	17430.766	56.103
2940	7.988	17749.907	56.212
2980	8.007	18069.816	56.320
3020	8.026	18390.484	56.427
3060	8.045	18711.900	56.533
3100	8.063	19034.054	56.637
3140	8.081	19356.937	56.741
3180	8.099	19680.539	56.843
3220	8.117	20004.850	56.945
3260	8.134	20329.861	57.045
3300	8.151	20655.562	57.144
3340	8.168	20981.944	57.243
3380	8.185	21308.998	57.340
3420	8.201	21636.714	57.436
3460	8.217	21965.084	57.532
3500	8.233	22294.099	57.626
3540	8.249	22623.749	57.720
3580	8.265	22954.026	57.813
3620	8.280	23284.921	57.905
3660	8.295	23616.426	57.996
3700	8.310	23948.532	58.086
3740	8.325	24281.230	58.175
3780	8.339	24614.514	58.264
3820	8.354	24948.373	58.352
3860	8.368	25282.800	58.439
3900	8.382	25617.787	58.525
3940	8.395	25953.327	58.611
3980	8.409	26289.410	58.696
4020	8.422	26626.030	58.780
4060	8.435	26963.179	58.863
4100	8.448	27300.849	58.946
4140	8.461	27639.032	59.028
4180	8.474	27977.722	59.110
4220	8.486	28316.910	59.190
4260	8.498	28656.591	59.271
4300	8.510	28996.756	59.350
4340	8.522	29337.398	59.429
4380	8.534	29678.511	59.507
4420	8.545	30020.087	59.585
4460	8.557	30362.120	59.662
4500	8.568	30704.604	59.738
4540	8.579	31047.531	59.814
4580	8.590	31390.895	59.889
4620	8.600	31734.689	59.964
4660	8.611	32078.908	60.038
4700	8.621	32423.544	60.112
4740	8.631	32768.592	60.185

T (R)	$\bar{c}_p$ (Btu/lbmol·R)	$\bar{h}°(T) - \bar{h}_f°(298)$ (Btu/lbmol)	$\bar{s}°(T)$ (Btu/lbmol·R)
4780	8.641	33114.046	60.258
4820	8.651	33459.900	60.330
4860	8.661	33806.147	60.401
4900	8.671	34152.782	60.472
5000	8.694	35021.028	60.648
5100	8.717	35891.575	60.820
5200	8.738	36764.338	60.990
5300	8.759	37639.236	61.156
5400	8.780	38516.190	61.320
5500	8.799	39395.127	61.481
5600	8.818	40275.975	61.640
5700	8.836	41158.665	61.796
5800	8.853	42043.132	61.950
5900	8.870	42929.314	62.102
6000	8.887	43817.152	62.251
6100	8.902	44706.590	62.398
6200	8.917	45597.575	62.543
6300	8.932	46490.058	62.686
6400	8.946	47383.990	62.826
6500	8.960	48279.329	62.965
6600	8.974	49176.034	63.102
6700	8.987	50074.066	63.237
6800	9.000	50973.391	63.370
6900	9.012	51873.975	63.502
7000	9.024	52775.791	63.632
7100	9.036	53678.810	63.760
7200	9.048	54583.010	63.886
7300	9.059	55488.370	64.011
7400	9.071	56394.871	64.134
7500	9.082	57302.499	64.256
7600	9.093	58211.239	64.377
7700	9.104	59121.083	64.495
7800	9.115	60032.023	64.613
7900	9.126	60944.054	64.729
8000	9.137	61857.174	64.844
8100	9.148	62771.385	64.958
8200	9.159	63686.688	65.070
8300	9.170	64603.090	65.181
8400	9.181	65520.599	65.291
8500	9.192	66439.226	65.400
8600	9.203	67358.983	65.507
8700	9.215	68279.887	65.614
8800	9.227	69201.956	65.719
8900	9.239	70125.210	65.823
9000	9.251	71049.672	65.927

Table BE.6   H₂O (Molecular Weight = 18.016, Enthalpy of Formation at 298 K = −103,974 Btu/lbmol)

T (R)	$\bar{c}_p$ (Btu/lbmol·R)	$\bar{h}°(T) - \bar{h}_f°(298)$ (Btu/lbmol)	$\bar{s}°(T)$ (Btu/lbmol·R)
9000	9.251	71049.672	65.927
300	7.586	−1846.132	40.548
320	7.626	−1694.010	41.039
340	7.666	−1541.084	41.502
360	7.704	−1387.385	41.942
380	7.740	−1232.941	42.359
400	7.776	−1077.777	42.757
420	7.810	−921.919	43.137
440	7.843	−765.390	43.501
460	7.875	−608.211	43.851
480	7.906	−450.404	44.187
500	7.936	−291.986	44.510
520	7.965	−132.976	44.822
536.67	7.989	0	45.073
540	7.994	26.611	45.123
560	8.021	186.759	45.414
580	8.048	347.456	45.696
600	8.075	508.690	45.969
620	8.101	670.448	46.234
640	8.126	832.722	46.492
660	8.152	995.503	46.743
680	8.176	1158.783	46.986
700	8.201	1322.556	47.224
720	8.225	1486.817	47.455
740	8.249	1651.560	47.681
760	8.273	1816.783	47.901
780	8.297	1982.484	48.116
800	8.321	2148.659	48.327
820	8.344	2315.310	48.532
840	8.368	2482.435	48.734
860	8.392	2650.036	48.931
880	8.416	2818.115	49.124
900	8.440	2986.673	49.313
920	8.464	3155.713	49.499
940	8.489	3325.240	49.682
960	8.513	3495.257	49.861
980	8.538	3665.769	50.036
1000	8.563	3836.782	50.209
1020	8.589	4008.302	50.379
1040	8.615	4180.334	50.546
1060	8.641	4352.885	50.710
1080	8.667	4525.963	50.872
1100	8.694	4699.574	51.031
1120	8.721	4873.728	51.188
1140	8.749	5048.431	51.343
1160	8.777	5223.691	51.495
1180	8.806	5399.518	51.645
1200	8.835	5575.921	51.794
1220	8.864	5752.906	51.940
1240	8.894	5930.485	52.084
1260	8.924	6108.665	52.227

T (R)	$\bar{c}_p$ (Btu/lbmol·R)	$\bar{h}°(T) - \bar{h}_f°(298)$ (Btu/lbmol)	$\bar{s}°(T)$ (Btu/lbmol·R)
1280	8.955	6287.455	52.368
1300	8.986	6466.864	52.507
1320	9.018	6646.902	52.644
1340	9.050	6827.577	52.780
1360	9.082	7008.897	52.914
1380	9.115	7190.871	53.047
1400	9.149	7373.508	53.179
1420	9.182	7556.815	53.309
1440	9.216	7740.800	53.437
1460	9.251	7925.470	53.565
1480	9.286	8110.833	53.691
1500	9.321	8296.896	53.816
1520	9.356	8483.665	53.939
1540	9.392	8671.145	54.062
1560	9.428	8859.342	54.183
1580	9.464	9048.261	54.304
1600	9.501	9237.907	54.423
1620	9.537	9428.282	54.541
1640	9.574	9619.390	54.658
1660	9.611	9811.233	54.775
1680	9.648	10003.813	54.890
1700	9.684	10197.132	55.004
1720	9.721	10391.188	55.118
1740	9.758	10585.982	55.230
1760	9.795	10781.511	55.342
1780	9.832	10977.775	55.453
1800	9.868	11174.768	55.563
1820	9.904	11372.490	55.672
1840	9.940	11570.929	55.781
1860	9.976	11770.084	55.888
1880	10.011	11969.948	55.995
1900	10.046	12170.518	56.101
1940	10.115	12573.748	56.311
1980	10.184	12979.735	56.519
2020	10.251	13388.435	56.723
2060	10.317	13799.809	56.925
2100	10.383	14213.816	57.124
2140	10.447	14630.415	57.320
2180	10.511	15049.569	57.514
2220	10.573	15471.238	57.706
2260	10.634	15895.383	57.895
2300	10.695	16321.966	58.082
2340	10.754	16750.951	58.267
2380	10.813	17182.299	58.450
2420	10.871	17615.975	58.631
2460	10.928	18051.942	58.809
2500	10.984	18490.164	58.986
2540	11.039	18930.607	59.161
2580	11.093	19373.235	59.334
2620	11.146	19818.013	59.505
2660	11.199	20264.909	59.674
2700	11.250	20713.889	59.842

(continued)

Table BE.6 (continued)

T (R)	$\bar{c}_p$ (Btu/lbmol·R)	$\bar{h}°(T) - \bar{h}_f°(298)$ (Btu/lbmol)	$\bar{s}°(T)$ (Btu/lbmol·R)
2740	11.301	21164.919	60.008
2780	11.351	21617.967	60.172
2820	11.400	22073.000	60.334
2860	11.449	22529.987	60.495
2900	11.497	22988.897	60.654
2940	11.543	23449.698	60.812
2980	11.590	23912.360	60.969
3020	11.635	24376.853	61.123
3060	11.680	24843.146	61.277
3100	11.724	25311.212	61.429
3140	11.767	25781.020	61.579
3180	11.809	26252.542	61.729
3220	11.851	26725.751	61.876
3260	11.892	27200.618	62.023
3300	11.933	27677.115	62.168
3340	11.972	28155.217	62.312
3380	12.012	28634.896	62.455
3420	12.050	29116.126	62.597
3460	12.088	29598.882	62.737
3500	12.125	30083.138	62.876
3540	12.162	30568.868	63.014
3580	12.198	31056.048	63.151
3620	12.233	31544.654	63.287
3660	12.268	32034.662	63.421
3700	12.302	32526.048	63.555
3740	12.335	33018.789	63.687
3780	12.368	33512.862	63.819
3820	12.401	34008.244	63.949
3860	12.433	34504.913	64.078
3900	12.464	35002.847	64.207
3940	12.495	35502.025	64.334
3980	12.525	36002.426	64.460
4020	12.555	36504.028	64.586
4060	12.584	37006.811	64.710
4100	12.613	37510.755	64.834
4140	12.641	38015.840	64.956
4180	12.669	38522.046	65.078
4220	12.696	39029.355	65.199
4260	12.723	39537.747	65.319
4300	12.750	40047.203	65.438
4340	12.776	40557.706	65.556
4380	12.801	41069.238	65.673
4420	12.826	41581.780	65.790
4460	12.851	42095.315	65.905
4500	12.875	42609.827	66.020
4540	12.899	43125.298	66.134
4580	12.922	43641.712	66.248
4620	12.945	44159.053	66.360
4660	12.968	44677.305	66.472
4700	12.990	45196.452	66.583
4740	13.012	45716.479	66.693
4780	13.033	46237.372	66.802

T (R)	$\bar{c}_p$ (Btu/lbmol·R)	$\bar{h}°(T) - \bar{h}_f°(298)$ (Btu/lbmol)	$\bar{s}°(T)$ (Btu/lbmol·R)
4820	13.054	46759.114	66.911
4860	13.075	47281.692	67.019
4900	13.095	47805.091	67.126
5000	13.145	49117.091	67.391
5100	13.192	50433.928	67.652
5200	13.237	51755.399	67.909
5300	13.281	53081.313	68.161
5400	13.323	54411.489	68.410
5500	13.363	55745.753	68.655
5600	13.401	57083.943	68.896
5700	13.438	58425.901	69.133
5800	13.474	59771.482	69.367
5900	13.508	61120.546	69.598
6000	13.541	62472.962	69.825
6100	13.572	63828.606	70.049
6200	13.603	65187.361	70.270
6300	13.632	66549.117	70.488
6400	13.661	67913.771	70.703
6500	13.688	69281.225	70.915
6600	13.715	70651.389	71.124
6700	13.741	72024.177	71.331
6800	13.766	73399.508	71.534
6900	13.790	74777.308	71.736
7000	13.814	76157.506	71.934
7100	13.837	77540.037	72.130
7200	13.859	78924.839	72.324
7300	13.881	80311.855	72.515
7400	13.902	81701.031	72.704
7500	13.923	83092.315	72.891
7600	13.944	84485.661	73.076
7700	13.964	85881.023	73.258
7800	13.983	87278.361	73.438
7900	14.002	88677.632	73.617
8000	14.021	90078.800	73.793
8100	14.040	91481.828	73.967
8200	14.058	92886.680	74.140
8300	14.075	94293.322	74.310
8400	14.093	95701.721	74.479
8500	14.110	97111.844	74.646
8600	14.127	98523.658	74.811
8700	14.143	99937.130	74.974
8800	14.159	101352.227	75.136
8900	14.175	102768.914	75.296
9000	14.190	104187.156	75.454

Table BE.7 N$_2$ (Molecular Weight = 28.013, Enthalpy of Formation at 298 K = 0 Btu/lbmol)

T (R)	$\bar{c}_p$ (Btu/lbmol·R)	$\bar{h}°(T) - \bar{h}°_f(298)$ (Btu/lbmol)	$\bar{s}°(T)$ (Btu/lbmol·R)
300	6.846	−1632.919	41.731
320	6.857	−1495.884	42.173
340	6.867	−1358.638	42.589
360	6.877	−1221.193	42.982
380	6.886	−1083.562	43.354
400	6.895	−945.756	43.708
420	6.903	−807.784	44.044
440	6.910	−669.656	44.366
460	6.918	−531.377	44.673
480	6.925	−392.955	44.967
500	6.931	−254.395	45.250
520	6.938	−115.701	45.522
536.67	6.943	0	45.741
540	6.944	23.123	45.784
560	6.951	162.075	46.037
580	6.957	301.153	46.281
600	6.963	440.357	46.517
620	6.970	579.686	46.745
640	6.976	719.141	46.967
660	6.982	858.725	47.181
680	6.989	998.439	47.390
700	6.996	1138.286	47.593
720	7.003	1278.270	47.790
740	7.010	1418.396	47.982
760	7.017	1558.667	48.169
780	7.025	1699.089	48.351
800	7.033	1839.669	48.529
820	7.041	1980.411	48.703
840	7.050	2121.324	48.873
860	7.059	2262.414	49.039
880	7.068	2403.688	49.201
900	7.078	2545.155	49.360
920	7.089	2686.821	49.516
940	7.099	2828.697	49.668
960	7.110	2970.789	49.818
980	7.122	3113.107	49.965
1000	7.134	3255.659	50.109
1020	7.146	3398.455	50.250
1040	7.159	3541.503	50.389
1060	7.172	3684.812	50.525
1080	7.186	3828.392	50.660
1100	7.200	3972.251	50.792
1120	7.215	4116.398	50.921
1140	7.230	4260.842	51.049
1160	7.245	4405.591	51.175
1180	7.261	4550.655	51.299
1200	7.278	4696.041	51.421
1220	7.294	4841.757	51.542
1240	7.311	4987.811	51.660
1260	7.329	5134.210	51.778
1280	7.347	5280.962	51.893

T (R)	$\bar{c}_p$ (Btu/lbmol·R)	$\bar{h}°(T) - \bar{h}_f°(298)$ (Btu/lbmol)	$\bar{s}°(T)$ (Btu/lbmol·R)
1300	7.365	5428.072	52.007
1320	7.383	5575.548	52.120
1340	7.402	5723.394	52.231
1360	7.421	5871.617	52.341
1380	7.440	6020.220	52.449
1400	7.459	6169.208	52.556
1420	7.479	6318.585	52.662
1440	7.498	6468.353	52.767
1460	7.518	6618.514	52.871
1480	7.538	6769.070	52.973
1500	7.558	6920.022	53.074
1520	7.577	7071.369	53.175
1540	7.597	7223.112	53.274
1560	7.617	7375.247	53.372
1580	7.636	7527.773	53.469
1600	7.655	7680.685	53.565
1620	7.674	7833.980	53.660
1640	7.693	7987.652	53.755
1660	7.711	8141.694	53.848
1680	7.729	8296.098	53.941
1700	7.747	8450.857	54.032
1720	7.764	8605.960	54.123
1740	7.780	8761.396	54.213
1760	7.796	8917.153	54.302
1780	7.811	9073.217	54.390
1800	7.825	9229.574	54.477
1820	7.839	9386.207	54.564
1840	7.852	9543.117	54.649
1860	7.866	9700.297	54.734
1880	7.879	9857.745	54.819
1900	7.892	10015.457	54.902
1940	7.918	10331.666	55.067
1980	7.944	10648.899	55.229
2020	7.968	10967.135	55.388
2060	7.993	11286.351	55.544
2100	8.016	11606.526	55.698
2140	8.039	11927.638	55.850
2180	8.062	12249.667	55.999
2220	8.084	12572.591	56.146
2260	8.106	12896.390	56.290
2300	8.127	13221.044	56.433
2340	8.148	13546.534	56.573
2380	8.168	13872.841	56.711
2420	8.187	14199.945	56.847
2460	8.207	14527.828	56.982
2500	8.226	14856.472	57.114
2540	8.244	15185.858	57.245
2580	8.262	15515.970	57.374
2620	8.279	15846.790	57.501
2660	8.296	16178.301	57.627
2700	8.313	16510.487	57.751

(continued)

Table BE.7 (continued)

T (R)	$\bar{c}_p$ (Btu/lbmol·R)	$\bar{h}°(T) - \bar{h}_f°(298)$ (Btu/lbmol)	$\bar{s}°(T)$ (Btu/lbmol·R)
2740	8.329	16843.331	57.873
2780	8.345	17176.818	57.994
2820	8.361	17510.931	58.113
2860	8.376	17845.655	58.231
2900	8.390	18180.976	58.348
2940	8.405	18516.879	58.463
2980	8.419	18853.349	58.576
3020	8.432	19190.372	58.689
3060	8.446	19527.934	58.800
3100	8.459	19866.022	58.909
3140	8.471	20204.623	59.018
3180	8.484	20543.723	59.125
3220	8.496	20883.311	59.231
3260	8.507	21223.372	59.336
3300	8.519	21563.897	59.440
3340	8.530	21904.871	59.543
3380	8.541	22246.285	59.644
3420	8.551	22588.126	59.745
3460	8.562	22930.384	59.845
3500	8.572	23273.048	59.943
3540	8.581	23616.107	60.040
3580	8.591	23959.551	60.137
3620	8.600	24303.369	60.232
3660	8.609	24647.553	60.327
3700	8.618	24992.093	60.421
3740	8.626	25336.979	60.513
3780	8.635	25682.202	60.605
3820	8.643	26027.754	60.696
3860	8.651	26373.625	60.786
3900	8.658	26719.808	60.875
3940	8.666	27066.294	60.964
3980	8.673	27413.076	61.051
4020	8.680	27760.145	61.138
4060	8.687	28107.495	61.224
4100	8.694	28455.117	61.309
4140	8.701	28803.005	61.394
4180	8.707	29151.153	61.477
4220	8.713	29499.552	61.560
4260	8.719	29848.198	61.643
4300	8.725	30197.083	61.724
4340	8.731	30546.201	61.805
4380	8.737	30895.548	61.885
4420	8.742	31245.116	61.964
4460	8.747	31594.901	62.043
4500	8.753	31944.897	62.121
4540	8.758	32295.099	62.199
4580	8.763	32645.502	62.276
4620	8.767	32996.101	62.352
4660	8.772	33346.892	62.428
4700	8.777	33697.871	62.503
4740	8.781	34049.031	62.577
4780	8.786	34400.371	62.651

T (R)	$\bar{c}_p$ (Btu/lbmol·R)	$\bar{h}°(T) - \bar{h}_f°(298)$ (Btu/lbmol)	$\bar{s}°(T)$ (Btu/lbmol·R)
4820	8.790	34751.886	62.724
4860	8.794	35103.571	62.797
4900	8.798	35455.424	62.869
5000	8.808	36335.764	63.047
5100	8.818	37217.076	63.221
5200	8.827	38099.314	63.392
5300	8.836	38982.436	63.561
5400	8.844	39866.408	63.726
5500	8.852	40751.195	63.888
5600	8.860	41636.771	64.048
5700	8.867	42523.110	64.205
5800	8.875	43410.194	64.359
5900	8.882	44298.002	64.511
6000	8.889	45186.522	64.660
6100	8.896	46075.741	64.807
6200	8.903	46965.648	64.952
6300	8.909	47856.236	65.094
6400	8.916	48747.499	65.235
6500	8.923	49639.430	65.373
6600	8.929	50532.026	65.509
6700	8.936	51425.284	65.644
6800	8.943	52319.201	65.776
6900	8.949	53213.773	65.907
7000	8.956	54108.998	66.035
7100	8.962	55004.871	66.162
7200	8.968	55901.389	66.288
7300	8.975	56798.545	66.412
7400	8.981	57696.332	66.534
7500	8.987	58594.742	66.654
7600	8.993	59493.764	66.773
7700	8.999	60393.384	66.891
7800	9.005	61293.586	67.007
7900	9.010	62194.352	67.122
8000	9.016	63095.659	67.235
8100	9.021	63997.481	67.347
8200	9.025	64899.789	67.458
8300	9.030	65802.549	67.567
8400	9.034	66705.721	67.676
8500	9.037	67609.264	67.783
8600	9.040	68513.128	67.888
8700	9.043	69417.260	67.993
8800	9.044	70321.600	68.096
8900	9.045	71226.082	68.198
9000	9.046	72130.635	68.299

Table BE.8 N (Molecular Weight = 14.007, Enthalpy of Formation at 298 K = 183,846 Btu/lbmol)

T (R)	$\bar{c}_p$ (Btu/lbmol·R)	$\bar{h}°(T) - \bar{h}_f°(298)$ (Btu/lbmol)	$\bar{s}°(T)$ (Btu/lbmol·R)
300	4.966	−1175.117	33.700
320	4.966	−1075.799	34.021
340	4.966	−976.485	34.322
360	4.966	−877.173	34.606
380	4.965	−777.865	34.874
400	4.965	−678.558	35.129
420	4.965	−579.254	35.371
440	4.965	−479.953	35.602
460	4.965	−380.653	35.823
480	4.965	−281.354	36.034
500	4.965	−182.057	36.237
520	4.965	−82.762	36.432
536.67	4.965	0	36.588
540	4.965	16.532	36.619
560	4.965	115.826	36.799
580	4.965	215.118	36.974
600	4.965	314.410	37.142
620	4.965	413.701	37.305
640	4.965	512.992	37.462
660	4.965	612.282	37.615
680	4.965	711.572	37.763
700	4.965	810.862	37.907
720	4.965	910.152	38.047
740	4.965	1009.441	38.183
760	4.965	1108.731	38.316
780	4.965	1208.020	38.445
800	4.965	1307.310	38.570
820	4.965	1406.600	38.693
840	4.965	1505.890	38.812
860	4.965	1605.180	38.929
880	4.965	1704.470	39.043
900	4.965	1803.761	39.155
920	4.965	1903.052	39.264
940	4.965	2002.343	39.371
960	4.965	2101.634	39.475
980	4.965	2200.926	39.578
1000	4.965	2300.218	39.678
1020	4.965	2399.510	39.776
1040	4.965	2498.802	39.873
1060	4.965	2598.095	39.967
1080	4.965	2697.387	40.060
1100	4.965	2796.680	40.151
1120	4.965	2895.973	40.241
1140	4.965	2995.266	40.329
1160	4.965	3094.560	40.415
1180	4.965	3193.853	40.500
1200	4.965	3293.146	40.583
1220	4.965	3392.440	40.665
1240	4.965	3491.733	40.746
1260	4.965	3591.027	40.825

T (R)	$\bar{c}_p$ (Btu/lbmol·R)	$\bar{h}°(T) - \bar{h}_f°(298)$ (Btu/lbmol)	$\bar{s}°(T)$ (Btu/lbmol·R)
1280	4.965	3690.320	40.904
1300	4.965	3789.613	40.981
1320	4.965	3888.906	41.056
1340	4.965	3988.199	41.131
1360	4.965	4087.492	41.205
1380	4.965	4186.785	41.277
1400	4.965	4286.077	41.349
1420	4.965	4385.370	41.419
1440	4.965	4484.662	41.488
1460	4.965	4583.954	41.557
1480	4.965	4683.246	41.624
1500	4.965	4782.537	41.691
1520	4.965	4881.828	41.757
1540	4.965	4981.119	41.822
1560	4.965	5080.410	41.886
1580	4.965	5179.701	41.949
1600	4.965	5278.991	42.011
1620	4.965	5378.281	42.073
1640	4.965	5477.572	42.134
1660	4.965	5576.862	42.194
1680	4.965	5676.152	42.254
1700	4.965	5775.442	42.312
1720	4.965	5874.732	42.370
1740	4.965	5974.023	42.428
1760	4.965	6073.313	42.485
1780	4.965	6172.605	42.541
1800	4.965	6271.896	42.596
1820	4.965	6371.206	42.651
1840	4.965	6470.504	42.705
1860	4.965	6569.805	42.759
1880	4.965	6669.110	42.812
1900	4.966	6768.419	42.865
1940	4.966	6967.044	42.968
1980	4.966	7165.681	43.069
2020	4.966	7364.327	43.169
2060	4.966	7562.980	43.266
2100	4.967	7761.639	43.362
2140	4.967	7960.303	43.455
2180	4.967	8158.970	43.547
2220	4.967	8357.640	43.638
2260	4.967	8556.310	43.726
2300	4.967	8754.981	43.814
2340	4.967	8953.650	43.899
2380	4.967	9152.317	43.983
2420	4.967	9350.981	44.066
2460	4.966	9549.641	44.148
2500	4.966	9748.297	44.228
2540	4.966	9946.947	44.306
2580	4.966	10145.592	44.384
2620	4.966	10344.230	44.460

(continued)

Table BE.8 (continued)

T (R)	$\bar{c}_p$ (Btu/lbmol·R)	$\bar{h}°(T) - \bar{h}_f°(298)$ (Btu/lbmol)	$\bar{s}°(T)$ (Btu/lbmol·R)
2660	4.966	10542.861	44.536
2700	4.966	10741.485	44.610
2740	4.965	10940.101	44.683
2780	4.965	11138.710	44.755
2820	4.965	11337.310	44.826
2860	4.965	11535.903	44.896
2900	4.965	11734.486	44.965
2940	4.964	11933.062	45.033
2980	4.964	12131.629	45.100
3020	4.964	12330.188	45.166
3060	4.964	12528.739	45.231
3100	4.964	12727.282	45.296
3140	4.963	12925.818	45.359
3180	4.963	13124.347	45.422
3220	4.963	13322.869	45.484
3260	4.963	13521.384	45.545
3300	4.963	13719.894	45.606
3340	4.963	13918.399	45.666
3380	4.962	14116.900	45.725
3420	4.962	14315.397	45.783
3460	4.962	14513.891	45.841
3500	4.962	14712.383	45.898
3540	4.962	14910.874	45.954
3580	4.962	15109.364	46.010
3620	4.962	15307.856	46.065
3660	4.962	15506.349	46.120
3700	4.962	15704.845	46.174
3740	4.963	15903.345	46.227
3780	4.963	16101.851	46.280
3820	4.963	16300.363	46.332
3860	4.963	16498.883	46.384
3900	4.963	16697.412	46.435
3940	4.964	16895.952	46.486
3980	4.964	17094.505	46.536
4020	4.964	17293.071	46.585
4060	4.965	17491.652	46.635
4100	4.965	17690.251	46.683
4140	4.966	17888.868	46.731
4180	4.966	18087.506	46.779
4220	4.967	18286.166	46.827
4260	4.967	18484.850	46.873
4300	4.968	18683.560	46.920
4340	4.969	18882.297	46.966
4380	4.970	19081.065	47.011
4420	4.970	19279.865	47.057
4460	4.971	19478.698	47.101
4500	4.972	19677.568	47.146
4540	4.973	19876.476	47.190
4580	4.974	20075.424	47.233
4620	4.975	20274.415	47.277
4660	4.977	20473.451	47.320
4700	4.978	20672.534	47.362

T (R)	$\bar{c}_p$ (Btu/lbmol·R)	$\bar{h}°(T) - \bar{h}_f°(298)$ (Btu/lbmol)	$\bar{s}°(T)$ (Btu/lbmol·R)
4740	4.979	20871.667	47.404
4780	4.980	21070.852	47.446
4820	4.982	21270.092	47.488
4860	4.983	21469.389	47.529
4900	4.985	21668.745	47.570
5000	4.989	22167.414	47.670
5100	4.993	22666.514	47.769
5200	4.998	23166.087	47.866
5300	5.004	23666.179	47.962
5400	5.010	24166.834	48.055
5500	5.016	24668.100	48.147
5600	5.023	25170.027	48.238
5700	5.030	25672.663	48.326
5800	5.038	26176.060	48.414
5900	5.046	26680.270	48.500
6000	5.055	27185.347	48.585
6100	5.065	27691.344	48.669
6200	5.075	28198.318	48.751
6300	5.085	28706.326	48.832
6400	5.097	29215.424	48.913
6500	5.108	29725.671	48.992
6600	5.121	30237.127	49.070
6700	5.134	30749.851	49.147
6800	5.147	31263.905	49.223
6900	5.162	31779.349	49.298
7000	5.176	32296.247	49.373
7100	5.192	32814.661	49.446
7200	5.208	33334.654	49.519
7300	5.225	33856.291	49.591
7400	5.242	34379.636	49.662
7500	5.260	34904.754	49.733
7600	5.279	35431.710	49.802
7700	5.298	35960.569	49.872
7800	5.318	36491.399	49.940
7900	5.339	37024.265	50.008
8000	5.360	37559.233	50.075
8100	5.382	38096.370	50.142
8200	5.405	38635.743	50.208
8300	5.429	39177.419	50.274
8400	5.453	39721.464	50.339
8500	5.477	40267.945	50.404
8600	5.503	40816.928	50.468
8700	5.529	41368.481	50.532
8800	5.555	41922.670	50.595
8900	5.583	42479.560	50.658
9000	5.611	43039.218	50.720

Table BE.9 NO (Molecular Weight = 30.006, Enthalpy of Formation at 298 K = 38,820 Btu/lbmol)

T (R)	$\bar{c}_p$ (Btu/lbmol·R)	$\bar{h}°(T) - \bar{h}_f°(298)$ (Btu/lbmol)	$\bar{s}°(T)$ (Btu/lbmol·R)
300	6.982	−1667.076	46.220
320	6.994	−1527.317	46.671
340	7.005	−1387.330	47.095
360	7.016	−1247.124	47.496
380	7.026	−1106.705	47.875
400	7.036	−966.082	48.236
420	7.046	−825.259	48.579
440	7.056	−684.242	48.907
460	7.065	−543.035	49.221
480	7.074	−401.641	49.522
500	7.084	−260.062	49.811
520	7.093	−118.299	50.089
536.67	7.100	0	50.313
540	7.102	23.647	50.357
560	7.111	165.777	50.615
580	7.120	308.091	50.865
600	7.130	450.592	51.107
620	7.139	593.282	51.341
640	7.149	736.164	51.567
660	7.159	879.242	51.788
680	7.169	1022.520	52.001
700	7.179	1166.003	52.209
720	7.190	1309.696	52.412
740	7.201	1453.605	52.609
760	7.212	1597.736	52.801
780	7.224	1742.095	52.989
800	7.236	1886.690	53.172
820	7.248	2031.529	53.350
840	7.261	2176.617	53.525
860	7.274	2321.964	53.696
880	7.287	2467.576	53.864
900	7.301	2613.464	54.028
920	7.316	2759.633	54.188
940	7.331	2906.094	54.346
960	7.346	3052.855	54.500
980	7.361	3199.924	54.652
1000	7.377	3347.309	54.801
1020	7.394	3495.019	54.947
1040	7.411	3643.063	55.091
1060	7.428	3791.449	55.232
1080	7.446	3940.185	55.371
1100	7.464	4089.279	55.508
1120	7.482	4238.739	55.642
1140	7.501	4388.573	55.775
1160	7.520	4538.787	55.906
1180	7.540	4689.388	56.034
1200	7.560	4840.384	56.161
1220	7.580	4991.780	56.286
1240	7.600	5143.582	56.410

T (R)	$\bar{c}_p$ (Btu/lbmol·R)	$\bar{h}°(T) - \bar{h}_f°(298)$ (Btu/lbmol)	$\bar{s}°(T)$ (Btu/lbmol·R)
1260	7.621	5295.795	56.532
1280	7.642	5448.424	56.652
1300	7.663	5601.474	56.770
1320	7.684	5754.948	56.888
1340	7.706	5908.850	57.003
1360	7.727	6063.180	57.118
1380	7.749	6217.943	57.231
1400	7.771	6373.137	57.342
1420	7.792	6528.765	57.453
1440	7.814	6684.825	57.562
1460	7.835	6841.316	57.670
1480	7.857	6998.236	57.776
1500	7.878	7155.581	57.882
1520	7.899	7313.349	57.987
1540	7.920	7471.533	58.090
1560	7.940	7630.129	58.192
1580	7.960	7789.129	58.294
1600	7.980	7948.525	58.394
1620	7.999	8108.308	58.493
1640	8.017	8268.468	58.591
1660	8.035	8428.993	58.689
1680	8.053	8589.871	58.785
1700	8.069	8751.088	58.880
1720	8.085	8912.629	58.975
1740	8.100	9074.477	59.068
1760	8.114	9236.614	59.161
1780	8.127	9399.022	59.253
1800	8.139	9561.679	59.344
1820	8.150	9724.563	59.434
1840	8.161	9887.677	59.523
1860	8.172	10051.013	59.611
1880	8.183	10214.569	59.699
1900	8.194	10378.342	59.785
1940	8.215	10706.532	59.956
1980	8.236	11035.561	60.124
2020	8.256	11365.412	60.289
2060	8.276	11696.065	60.451
2100	8.296	12027.501	60.610
2140	8.314	12359.702	60.767
2180	8.333	12692.650	60.921
2220	8.351	13026.327	61.073
2260	8.369	13360.715	61.222
2300	8.386	13695.799	61.369
2340	8.402	14031.560	61.514
2380	8.419	14367.983	61.656
2420	8.435	14705.051	61.797
2460	8.450	15042.749	61.935
2500	8.465	15381.061	62.072
2540	8.480	15719.972	62.206
2580	8.495	16059.466	62.339
2620	8.509	16399.530	62.470

(continued)

Table BE.9 (continued)

T (R)	$\bar{c}_p$ (Btu/lbmol·R)	$\bar{h}°(T) - \bar{h}_f°(298)$ (Btu/lbmol)	$\bar{s}°(T)$ (Btu/lbmol·R)
2660	8.522	16740.149	62.599
2700	8.536	17081.308	62.726
2740	8.549	17422.995	62.852
2780	8.561	17765.195	62.976
2820	8.574	18107.895	63.098
2860	8.586	18451.083	63.219
2900	8.597	18794.746	63.338
2940	8.609	19138.871	63.456
2980	8.620	19483.447	63.572
3020	8.631	19828.461	63.687
3060	8.641	20173.902	63.801
3100	8.652	20519.758	63.913
3140	8.662	20866.019	64.024
3180	8.671	21212.673	64.134
3220	8.681	21559.711	64.242
3260	8.690	21907.121	64.350
3300	8.699	22254.894	64.456
3340	8.708	22603.019	64.561
3380	8.716	22951.488	64.664
3420	8.724	23300.291	64.767
3460	8.732	23649.419	64.868
3500	8.740	23998.862	64.969
3540	8.748	24348.613	65.068
3580	8.755	24698.662	65.166
3620	8.762	25049.002	65.264
3660	8.769	25399.624	65.360
3700	8.776	25750.521	65.455
3740	8.782	26101.685	65.550
3780	8.789	26453.109	65.643
3820	8.795	26804.786	65.736
3860	8.801	27156.708	65.828
3900	8.807	27508.869	65.918
3940	8.813	27861.262	66.008
3980	8.818	28213.882	66.097
4020	8.824	28566.721	66.185
4060	8.829	28919.774	66.273
4100	8.834	29273.035	66.359
4140	8.839	29626.498	66.445
4180	8.844	29980.159	66.530
4220	8.849	30334.011	66.614
4260	8.853	30688.050	66.698
4300	8.858	31042.270	66.781
4340	8.862	31396.668	66.863
4380	8.866	31751.238	66.944
4420	8.871	32105.975	67.025
4460	8.875	32460.877	67.105
4500	8.879	32815.939	67.184
4540	8.882	33171.156	67.263
4580	8.886	33526.525	67.340
4620	8.890	33882.042	67.418
4660	8.893	34237.704	67.494
4700	8.897	34593.508	67.570

T (R)	$\bar{c}_p$ (Btu/lbmol·R)	$\bar{h}°(T) - \bar{h}_f°(298)$ (Btu/lbmol)	$\bar{s}°(T)$ (Btu/lbmol·R)
4740	8.900	34949.450	67.646
4780	8.904	35305.527	67.721
4820	8.907	35661.736	67.795
4860	8.910	36018.075	67.868
4900	8.913	36374.541	67.942
5000	8.921	37266.243	68.122
5100	8.928	38158.685	68.298
5200	8.935	39051.835	68.472
5300	8.942	39945.664	68.642
5400	8.948	40840.149	68.809
5500	8.954	41735.270	68.974
5600	8.960	42631.009	69.135
5700	8.966	43527.352	69.294
5800	8.972	44424.289	69.450
5900	8.978	45321.809	69.603
6000	8.984	46219.908	69.754
6100	8.990	47118.580	69.902
6200	8.995	48017.822	70.049
6300	9.001	48917.633	70.193
6400	9.007	49818.012	70.334
6500	9.012	50718.959	70.474
6600	9.018	51620.476	70.612
6700	9.024	52522.562	70.747
6800	9.029	53425.220	70.881
6900	9.035	54328.449	71.013
7000	9.041	55232.249	71.143
7100	9.047	56136.620	71.271
7200	9.052	57041.560	71.398
7300	9.058	57947.063	71.523
7400	9.063	58853.126	71.646
7500	9.069	59759.740	71.768
7600	9.074	60666.895	71.888
7700	9.079	61574.579	72.007
7800	9.085	62482.776	72.124
7900	9.089	63391.469	72.240
8000	9.094	64300.633	72.354
8100	9.098	65210.244	72.467
8200	9.102	66120.272	72.579
8300	9.106	67030.682	72.689
8400	9.109	67941.435	72.798
8500	9.112	68852.487	72.906
8600	9.114	69763.789	73.012
8700	9.116	70675.287	73.118
8800	9.117	71586.921	73.222
8900	9.117	72498.623	73.325
9000	9.117	73410.321	73.427

Table BE.10 NO$_2$ (Molecular Weight = 30.006, Enthalpy of Formation at 298 K = 14,229.5 Btu/lbmol)

T (R)	$\bar{c}_p$ (Btu/lbmol·R)	$\bar{h}°(T) - \bar{h}_f°(298)$ (Btu/lbmol)	$\bar{s}°(T)$ (Btu/lbmol·R)
300	7.506	−1936.740	52.582
320	7.629	−1785.389	53.070
340	7.749	−1631.609	53.536
360	7.867	−1475.451	53.983
380	7.982	−1316.964	54.411
400	8.095	−1156.195	54.823
420	8.205	−993.190	55.221
440	8.314	−827.995	55.605
460	8.420	−660.653	55.977
480	8.524	−491.205	56.338
500	8.627	−319.694	56.688
520	8.727	−146.160	57.028
536.67	8.809	0	57.305
540	8.825	29.360	57.359
560	8.922	206.828	57.682
580	9.016	386.207	57.997
600	9.109	567.461	58.304
620	9.200	750.557	58.604
640	9.290	935.462	58.898
660	9.378	1122.142	59.185
680	9.464	1310.567	59.466
700	9.549	1500.705	59.742
720	9.633	1692.528	60.012
740	9.715	1886.006	60.277
760	9.796	2081.111	60.537
780	9.875	2277.817	60.792
800	9.953	2476.096	61.043
820	10.030	2675.923	61.290
840	10.105	2877.273	61.533
860	10.180	3080.120	61.771
880	10.253	3284.442	62.006
900	10.325	3490.215	62.238
920	10.395	3697.416	62.465
940	10.465	3906.022	62.690
960	10.534	4116.013	62.911
980	10.601	4327.365	63.128
1000	10.668	4540.059	63.343
1020	10.733	4754.074	63.555
1040	10.798	4969.389	63.764
1060	10.862	5185.984	63.971
1080	10.924	5403.840	64.174
1100	10.986	5622.936	64.375
1120	11.046	5843.254	64.574
1140	11.106	6064.775	64.770
1160	11.165	6287.478	64.963
1180	11.222	6511.346	65.155
1200	11.279	6736.360	65.344
1220	11.335	6962.499	65.531
1240	11.390	7189.746	65.715
1260	11.444	7418.082	65.898

T (R)	$\bar{c}_p$ (Btu/lbmol·R)	$\bar{h}°(T) - \bar{h}°_f(298)$ (Btu/lbmol)	$\bar{s}°(T)$ (Btu/lbmol·R)
1280	11.497	7647.487	66.079
1300	11.549	7877.943	66.257
1320	11.600	8109.429	66.434
1340	11.650	8341.927	66.609
1360	11.699	8575.418	66.782
1380	11.747	8809.880	66.953
1400	11.794	9045.294	67.122
1420	11.840	9281.639	67.290
1440	11.885	9518.895	67.456
1460	11.929	9757.040	67.620
1480	11.972	9996.052	67.783
1500	12.014	10235.911	67.944
1520	12.054	10476.592	68.103
1540	12.094	10718.072	68.261
1560	12.132	10960.330	68.417
1580	12.169	11203.339	68.572
1600	12.205	11447.075	68.725
1620	12.239	11691.513	68.877
1640	12.272	11936.627	69.028
1660	12.304	12182.389	69.177
1680	12.334	12428.772	69.324
1700	12.363	12675.748	69.470
1720	12.391	12923.287	69.615
1740	12.417	13171.360	69.758
1760	12.441	13419.935	69.900
1780	12.464	13668.981	70.041
1800	12.485	13918.464	70.181
1820	12.505	14168.360	70.319
1840	12.524	14418.651	70.455
1860	12.544	14669.335	70.591
1880	12.563	14920.408	70.725
1900	12.582	15171.864	70.858
1940	12.620	15675.907	71.121
1980	12.656	16181.423	71.379
2020	12.691	16688.375	71.632
2060	12.726	17196.724	71.881
2100	12.759	17706.432	72.126
2140	12.792	18217.463	72.367
2180	12.824	18729.780	72.605
2220	12.855	19243.349	72.838
2260	12.885	19758.135	73.068
2300	12.914	20274.103	73.294
2340	12.942	20791.221	73.517
2380	12.970	21309.455	73.737
2420	12.996	21828.773	73.953
2460	13.022	22349.145	74.166
2500	13.047	22870.540	74.377
2540	13.072	23392.926	74.584
2580	13.096	23916.276	74.788
2620	13.119	24440.560	74.990
2660	13.141	24965.749	75.189

(continued)

Table BE.10 (continued)

T (R)	$\bar{c}_p$ (Btu/lbmol·R)	$\bar{h}°(T) - \bar{h}_f°(298)$ (Btu/lbmol)	$\bar{s}°(T)$ (Btu/lbmol·R)
2700	13.163	25491.817	75.385
2740	13.183	26018.736	75.579
2780	13.204	26546.479	75.770
2820	13.223	27075.022	75.959
2860	13.242	27604.338	76.145
2900	13.261	28134.404	76.329
2940	13.279	28665.194	76.511
2980	13.296	29196.686	76.691
3020	13.313	29728.856	76.868
3060	13.329	30261.682	77.043
3100	13.344	30795.142	77.217
3140	13.359	31329.215	77.388
3180	13.374	31863.880	77.557
3220	13.388	32399.117	77.724
3260	13.402	32934.906	77.890
3300	13.415	33471.228	78.053
3340	13.427	34008.063	78.215
3380	13.439	34545.394	78.375
3420	13.451	35083.204	78.533
3460	13.462	35621.474	78.689
3500	13.473	36160.188	78.844
3540	13.484	36699.329	78.997
3580	13.494	37238.883	79.149
3620	13.504	37778.833	79.299
3660	13.513	38319.164	79.447
3700	13.522	38859.863	79.594
3740	13.531	39400.915	79.740
3780	13.539	39942.306	79.884
3820	13.547	40484.023	80.026
3860	13.555	41026.054	80.167
3900	13.562	41568.387	80.307
3940	13.569	42111.008	80.446
3980	13.576	42653.908	80.583
4020	13.582	43197.074	80.719
4060	13.589	43740.496	80.853
4100	13.595	44284.164	80.986
4140	13.601	44828.067	81.118
4180	13.606	45372.197	81.249
4220	13.611	45916.543	81.379
4260	13.616	46461.097	81.507
4300	13.621	47005.851	81.634
4340	13.626	47550.796	81.761
4380	13.631	48095.925	81.886
4420	13.635	48641.230	82.010
4460	13.639	49186.703	82.132
4500	13.643	49732.338	82.254
4540	13.647	50278.129	82.375
4580	13.650	50824.068	82.495
4620	13.654	51370.151	82.613
4660	13.657	51916.372	82.731
4700	13.661	52462.724	82.848
4740	13.664	53009.204	82.964

T (R)	$\bar{c}_p$ (Btu/lbmol·R)	$\bar{h}°(T) - \bar{h}_f°(298)$ (Btu/lbmol)	$\bar{s}°(T)$ (Btu/lbmol·R)
4780	13.667	53555.806	83.078
4820	13.670	54102.526	83.192
4860	13.672	54649.359	83.305
4900	13.675	55196.302	83.417
5000	13.681	56564.113	83.694
5100	13.687	57932.533	83.965
5200	13.693	59301.518	84.231
5300	13.698	60671.032	84.491
5400	13.703	62041.047	84.748
5500	13.707	63411.542	84.999
5600	13.712	64782.502	85.246
5700	13.717	66153.918	85.489
5800	13.721	67525.786	85.727
5900	13.726	68898.105	85.962
6000	13.730	70270.880	86.193
6100	13.735	71644.117	86.420
6200	13.740	73017.827	86.643
6300	13.744	74392.020	86.863
6400	13.749	75766.708	87.079
6500	13.755	77141.906	87.293
6600	13.760	78517.625	87.503
6700	13.765	79893.879	87.710
6800	13.771	81270.679	87.914
6900	13.776	82648.033	88.115
7000	13.782	84025.948	88.313
7100	13.788	85404.427	88.509
7200	13.793	86783.470	88.701
7300	13.799	88163.072	88.892
7400	13.804	89543.223	89.079
7500	13.810	90923.907	89.265
7600	13.814	92305.101	89.448
7700	13.819	93686.777	89.628
7800	13.823	95068.898	89.807
7900	13.827	96451.419	89.983
8000	13.830	97834.285	90.157
8100	13.833	99217.433	90.329
8200	13.834	100600.789	90.498
8300	13.835	101984.270	90.666
8400	13.835	103367.778	90.832
8500	13.834	104751.207	90.995
8600	13.831	106134.435	91.157
8700	13.827	107517.329	91.317
8800	13.821	108899.741	91.475
8900	13.814	110281.507	91.631
9000	13.805	111662.451	91.786

Table BE.11 O_2 (Molecular Weight = 31.999, Enthalpy of Formation at 298 K = 0 Btu/lbmol)

T (R)	$\bar{c}_p$ (Btu/lbmol·R)	$\bar{h}°(T) - \bar{h}_f°(298)$ (Btu/lbmol)	$\bar{s}°(T)$ (Btu/lbmol·R)
300	6.733	−1625.167	44.987
320	6.755	−1490.290	45.422
340	6.778	−1354.959	45.833
360	6.801	−1219.175	46.221
380	6.823	−1082.938	46.589
400	6.846	−946.248	46.939
420	6.869	−809.104	47.274
440	6.891	−671.506	47.594
460	6.914	−533.454	47.901
480	6.937	−394.945	48.196
500	6.960	−255.981	48.479
520	6.983	−116.559	48.753
536.67	7.002	0	48.973
540	7.006	23.322	49.017
560	7.029	163.663	49.272
580	7.052	304.465	49.519
600	7.075	445.729	49.758
620	7.098	587.458	49.991
640	7.121	729.652	50.216
660	7.145	872.314	50.436
680	7.168	1015.444	50.650
700	7.192	1159.045	50.858
720	7.216	1303.118	51.061
740	7.239	1447.665	51.259
760	7.263	1592.686	51.452
780	7.287	1738.185	51.641
800	7.311	1884.161	51.826
820	7.335	2030.617	52.007
840	7.359	2177.554	52.184
860	7.383	2324.972	52.357
880	7.407	2472.874	52.527
900	7.431	2621.259	52.694
920	7.456	2770.130	52.857
940	7.480	2919.486	53.018
960	7.504	3069.329	53.176
980	7.529	3219.658	53.331
1000	7.553	3370.473	53.483
1020	7.577	3521.776	53.633
1040	7.602	3673.565	53.780
1060	7.626	3825.840	53.925
1080	7.650	3978.601	54.068
1100	7.674	4131.846	54.209
1120	7.699	4285.574	54.347
1140	7.723	4439.785	54.484
1160	7.747	4594.476	54.618
1180	7.770	4749.645	54.751
1200	7.794	4905.290	54.882
1220	7.818	5061.409	55.011
1240	7.841	5217.997	55.138
1260	7.864	5375.053	55.264

T (R)	$\bar{c}_p$ (Btu/lbmol·R)	$\bar{h}°(T) - \bar{h}_f°(298)$ (Btu/lbmol)	$\bar{s}°(T)$ (Btu/lbmol·R)
1280	7.888	5532.572	55.388
1300	7.910	5690.550	55.510
1320	7.933	5848.984	55.631
1340	7.955	6007.867	55.751
1360	7.977	6167.195	55.869
1380	7.999	6326.961	55.985
1400	8.021	6487.161	56.100
1420	8.042	6647.788	56.214
1440	8.063	6808.833	56.327
1460	8.083	6970.290	56.438
1480	8.103	7132.152	56.548
1500	8.123	7294.408	56.657
1520	8.142	7457.051	56.765
1540	8.160	7620.071	56.872
1560	8.178	7783.458	56.977
1580	8.196	7947.200	57.081
1600	8.213	8111.288	57.184
1620	8.229	8275.709	57.287
1640	8.245	8440.451	57.388
1660	8.260	8605.500	57.488
1680	8.274	8770.844	57.587
1700	8.288	8936.468	57.685
1720	8.301	9102.358	57.782
1740	8.313	9268.497	57.878
1760	8.324	9434.870	57.973
1780	8.335	9601.460	58.067
1800	8.344	9768.249	58.160
1820	8.353	9935.224	58.252
1840	8.362	10102.377	58.344
1860	8.371	10269.710	58.434
1880	8.380	10437.222	58.524
1900	8.389	10604.912	58.612
1940	8.407	10940.824	58.787
1980	8.424	11277.438	58.959
2020	8.442	11614.749	59.128
2060	8.459	11952.752	59.294
2100	8.476	12291.440	59.456
2140	8.493	12630.808	59.616
2180	8.509	12970.850	59.774
2220	8.526	13311.560	59.929
2260	8.543	13652.934	60.081
2300	8.559	13994.964	60.231
2340	8.575	14337.646	60.379
2380	8.591	14680.975	60.524
2420	8.607	15024.945	60.668
2460	8.623	15369.551	60.809
2500	8.639	15714.788	60.948
2540	8.654	16060.650	61.085
2580	8.670	16407.133	61.221
2620	8.685	16754.231	61.354
2660	8.700	17101.939	61.486

(*continued*)

Table BE.11 (continued)

T (R)	$\bar{c}_p$ (Btu/lbmol·R)	$\bar{h}°(T) - \bar{h}_f°(298)$ (Btu/lbmol)	$\bar{s}°(T)$ (Btu/lbmol·R)
2700	8.715	17450.254	61.616
2740	8.730	17799.169	61.744
2780	8.745	18148.680	61.871
2820	8.760	18498.783	61.996
2860	8.775	18849.472	62.119
2900	8.789	19200.743	62.241
2940	8.803	19552.591	62.362
2980	8.818	19905.013	62.481
3020	8.832	20258.003	62.599
3060	8.846	20611.557	62.715
3100	8.860	20965.671	62.830
3140	8.874	21320.340	62.944
3180	8.887	21675.560	63.056
3220	8.901	22031.328	63.167
3260	8.915	22387.638	63.277
3300	8.928	22744.487	63.386
3340	8.941	23101.871	63.494
3380	8.955	23459.786	63.600
3420	8.968	23818.227	63.706
3460	8.981	24177.191	63.810
3500	8.994	24536.675	63.913
3540	9.006	24896.674	64.015
3580	9.019	25257.184	64.117
3620	9.032	25618.202	64.217
3660	9.044	25979.724	64.316
3700	9.057	26341.747	64.415
3740	9.069	26704.267	64.512
3780	9.082	27067.280	64.609
3820	9.094	27430.783	64.704
3860	9.106	27794.773	64.799
3900	9.118	28159.246	64.893
3940	9.130	28524.199	64.986
3980	9.142	28889.628	65.078
4020	9.154	29255.531	65.170
4060	9.165	29621.903	65.261
4100	9.177	29988.742	65.351
4140	9.188	30356.045	65.440
4180	9.200	30723.808	65.528
4220	9.211	31092.029	65.616
4260	9.223	31460.704	65.703
4300	9.234	31829.831	65.789
4340	9.245	32199.405	65.875
4380	9.256	32569.426	65.959
4420	9.267	32939.889	66.044
4460	9.278	33310.792	66.127
4500	9.289	33682.131	66.210
4540	9.300	34053.905	66.292
4580	9.311	34426.110	66.374
4620	9.321	34798.744	66.455
4660	9.332	35171.804	66.535
4700	9.342	35545.287	66.615
4740	9.353	35919.191	66.694

T (R)	$\bar{c}_p$ (Btu/lbmol·R)	$\bar{h}°(T) - \bar{h}_f°(298)$ (Btu/lbmol)	$\bar{s}°(T)$ (Btu/lbmol·R)
4780	9.363	36293.513	66.773
4820	9.374	36668.250	66.851
4860	9.384	37043.400	66.929
4900	9.394	37418.961	67.006
5000	9.419	38359.642	67.196
5100	9.444	39302.835	67.382
5200	9.469	40248.503	67.566
5300	9.493	41196.611	67.747
5400	9.517	42147.122	67.924
5500	9.541	43100.003	68.099
5600	9.564	44055.221	68.271
5700	9.587	45012.742	68.441
5800	9.609	45972.537	68.608
5900	9.632	46934.572	68.772
6000	9.653	47898.819	68.934
6100	9.675	48865.246	69.094
6200	9.697	49833.825	69.251
6300	9.718	50804.528	69.407
6400	9.738	51777.324	69.560
6500	9.759	52752.187	69.711
6600	9.779	53729.089	69.860
6700	9.799	54708.003	70.007
6800	9.819	55688.900	70.153
6900	9.838	56671.754	70.296
7000	9.857	57656.539	70.438
7100	9.876	58643.227	70.578
7200	9.895	59631.792	70.716
7300	9.913	60622.206	70.853
7400	9.931	61614.442	70.988
7500	9.949	62608.475	71.121
7600	9.967	63604.275	71.253
7700	9.984	64601.817	71.383
7800	10.001	65601.071	71.512
7900	10.018	66602.010	71.640
8000	10.034	67604.605	71.766
8100	10.050	68608.827	71.891
8200	10.066	69614.647	72.014
8300	10.082	70622.034	72.136
8400	10.097	71630.958	72.257
8500	10.112	72641.388	72.377
8600	10.126	73653.291	72.495
8700	10.141	74666.635	72.612
8800	10.154	75681.386	72.728
8900	10.168	76697.509	72.843
9000	10.181	77714.970	72.957

Table BE.12   O (Molecular Weight = 16.000, Enthalpy of Formation at 298 K = 107,134 Btu/lbmol)

T (R)	$\bar{c}_p$ (Btu/lbmol·R)	$\bar{h}°(T) - \bar{h}_f°(298)$ (Btu/lbmol)	$\bar{s}°(T)$ (Btu/lbmol·R)
300	5.428	−1259.354	35.340
320	5.408	−1150.996	35.690
340	5.388	−1043.044	36.017
360	5.369	−935.482	36.325
380	5.350	−828.296	36.614
400	5.333	−721.469	36.888
420	5.316	−614.988	37.148
440	5.299	−508.838	37.395
460	5.284	−403.005	37.630
480	5.269	−297.476	37.855
500	5.255	−192.239	38.070
520	5.241	−87.280	38.276
536.67	5.230	0	38.441
540	5.228	17.413	38.473
560	5.216	121.852	38.663
580	5.204	226.048	38.846
600	5.193	330.012	39.022
620	5.182	433.755	39.192
640	5.172	537.287	39.356
660	5.162	640.619	39.515
680	5.152	743.760	39.669
700	5.144	846.719	39.819
720	5.135	949.506	39.963
740	5.127	1052.129	40.104
760	5.120	1154.597	40.241
780	5.112	1256.917	40.374
800	5.106	1359.098	40.503
820	5.099	1461.147	40.629
840	5.093	1563.070	40.752
860	5.087	1664.876	40.871
880	5.082	1766.570	40.988
900	5.077	1868.159	41.103
920	5.072	1969.648	41.214
940	5.068	2071.044	41.323
960	5.063	2172.353	41.430
980	5.059	2273.578	41.534
1000	5.056	2374.726	41.636
1020	5.052	2475.801	41.736
1040	5.049	2576.807	41.834
1060	5.046	2677.749	41.931
1080	5.043	2778.631	42.025
1100	5.040	2879.456	42.117
1120	5.037	2980.228	42.208
1140	5.035	3080.951	42.297
1160	5.033	3181.628	42.385
1180	5.031	3282.261	42.471
1200	5.029	3382.854	42.555
1220	5.027	3483.409	42.638
1240	5.025	3583.928	42.720
1260	5.024	3684.413	42.801
1280	5.022	3784.867	42.880

T (R)	$\overline{c}_p$ (Btu/lbmol·R)	$\overline{h}°(T) - \overline{h}_f°(298)$ (Btu/lbmol)	$\overline{s}°(T)$ (Btu/lbmol·R)
1300	5.021	3885.292	42.958
1320	5.019	3985.689	43.034
1340	5.018	4086.060	43.110
1360	5.017	4186.405	43.184
1380	5.016	4286.727	43.257
1400	5.014	4387.027	43.329
1420	5.013	4487.305	43.400
1440	5.012	4587.563	43.471
1460	5.011	4687.800	43.540
1480	5.010	4788.018	43.608
1500	5.010	4888.218	43.675
1520	5.009	4988.398	43.741
1540	5.008	5088.561	43.807
1560	5.007	5188.706	43.872
1580	5.006	5288.833	43.935
1600	5.005	5388.942	43.998
1620	5.004	5489.034	44.060
1640	5.003	5589.107	44.122
1660	5.002	5689.162	44.183
1680	5.001	5789.199	44.242
1700	5.000	5889.217	44.302
1720	5.000	5989.217	44.360
1740	4.999	6089.196	44.418
1760	4.998	6189.156	44.475
1780	4.996	6289.095	44.531
1800	4.995	6389.014	44.587
1820	4.995	6488.923	44.642
1840	4.994	6588.817	44.697
1860	4.994	6688.701	44.751
1880	4.994	6788.576	44.804
1900	4.993	6888.441	44.857
1940	4.992	7088.145	44.961
1980	4.991	7287.812	45.063
2020	4.990	7487.443	45.163
2060	4.990	7687.040	45.261
2100	4.989	7886.602	45.357
2140	4.988	8086.131	45.451
2180	4.987	8285.627	45.543
2220	4.986	8485.091	45.634
2260	4.985	8684.524	45.723
2300	4.985	8883.926	45.810
2340	4.984	9083.298	45.896
2380	4.983	9282.642	45.981
2420	4.983	9481.957	46.064
2460	4.982	9681.246	46.146
2500	4.981	9880.507	46.226
2540	4.981	10079.743	46.305
2580	4.980	10278.955	46.383
2620	4.979	10478.142	46.459
2660	4.979	10677.307	46.535

(continued)

Table BE.12 (continued)

T (R)	$\bar{c}_p$ (Btu/lbmol·R)	$\bar{h}°(T) - \bar{h}_f°(298)$ (Btu/lbmol)	$\bar{s}°(T)$ (Btu/lbmol·R)
2700	4.978	10876.449	46.609
2740	4.978	11075.570	46.682
2780	4.977	11274.671	46.755
2820	4.977	11473.752	46.826
2860	4.976	11672.815	46.896
2900	4.976	11871.860	46.965
2940	4.976	12070.888	47.033
2980	4.975	12269.901	47.100
3020	4.975	12468.900	47.167
3060	4.974	12667.884	47.232
3100	4.974	12866.856	47.297
3140	4.974	13065.816	47.360
3180	4.974	13264.765	47.423
3220	4.973	13463.704	47.486
3260	4.973	13662.635	47.547
3300	4.973	13861.558	47.608
3340	4.973	14060.474	47.668
3380	4.973	14259.385	47.727
3420	4.973	14458.291	47.785
3460	4.973	14657.193	47.843
3500	4.973	14856.093	47.900
3540	4.972	15054.991	47.957
3580	4.972	15253.889	48.013
3620	4.973	15452.787	48.068
3660	4.973	15651.687	48.122
3700	4.973	15850.590	48.177
3740	4.973	16049.497	48.230
3780	4.973	16248.409	48.283
3820	4.973	16447.327	48.335
3860	4.973	16646.252	48.387
3900	4.973	16845.186	48.438
3940	4.974	17044.129	48.489
3980	4.974	17243.082	48.539
4020	4.974	17442.047	48.589
4060	4.975	17641.025	48.638
4100	4.975	17840.017	48.687
4140	4.975	18039.023	48.735
4180	4.976	18238.046	48.783
4220	4.976	18437.086	48.831
4260	4.977	18636.145	48.878
4300	4.977	18835.223	48.924
4340	4.978	19034.322	48.970
4380	4.978	19233.442	49.016
4420	4.979	19432.586	49.061
4460	4.980	19631.753	49.106
4500	4.980	19830.946	49.150
4540	4.981	20030.166	49.194
4580	4.982	20229.413	49.238
4620	4.982	20428.688	49.281
4660	4.983	20627.994	49.324
4700	4.984	20827.330	49.367

T (R)	$\bar{c}_p$ (Btu/lbmol·R)	$\bar{h}°(T) - \bar{h}_f°(298)$ (Btu/lbmol)	$\bar{s}°(T)$ (Btu/lbmol·R)
4740	4.985	21026.699	49.409
4780	4.986	21226.101	49.451
4820	4.986	21425.538	49.493
4860	4.987	21625.010	49.534
4900	4.988	21824.519	49.575
5000	4.991	22323.460	49.676
5100	4.993	22822.655	49.774
5200	4.996	23322.124	49.871
5300	4.999	23821.883	49.967
5400	5.002	24321.951	50.060
5500	5.006	24822.344	50.152
5600	5.009	25323.080	50.242
5700	5.013	25824.178	50.331
5800	5.017	26325.652	50.418
5900	5.021	26827.522	50.504
6000	5.025	27329.802	50.588
6100	5.029	27832.510	50.671
6200	5.034	28335.661	50.753
6300	5.038	28839.272	50.834
6400	5.043	29343.358	50.913
6500	5.048	29847.935	50.991
6600	5.053	30353.017	51.069
6700	5.059	30858.619	51.145
6800	5.064	31364.754	51.220
6900	5.070	31871.438	51.294
7000	5.075	32378.683	51.366
7100	5.081	32886.502	51.439
7200	5.087	33394.907	51.510
7300	5.093	33903.911	51.580
7400	5.099	34413.525	51.649
7500	5.106	34923.760	51.718
7600	5.112	35434.626	51.785
7700	5.118	35946.134	51.852
7800	5.125	36458.292	51.918
7900	5.132	36971.110	51.984
8000	5.138	37484.595	52.048
8100	5.145	37998.754	52.112
8200	5.152	38513.595	52.175
8300	5.159	39029.124	52.238
8400	5.166	39545.347	52.300
8500	5.173	40062.266	52.361
8600	5.180	40579.888	52.421
8700	5.187	41098.215	52.481
8800	5.194	41617.249	52.541
8900	5.201	42136.992	52.599
9000	5.208	42657.445	52.657

Table BE.13   C(s) (Graphite, Molecular Weight = 12.011, Enthalpy of Formation at 298 K = 0 Btu/lbmol)

T (R)	$\bar{c}_p$ (Btu/lbmol·R)	$\bar{h}°(T) - \bar{h}_f°(298)$ (Btu/lbmol)	$\bar{s}°(T)$ (Btu/lbmol·R)
300	0.754	−335.447	0.582
320	0.873	−319.175	0.634
340	0.991	−300.532	0.691
360	1.106	−279.563	0.750
380	1.219	−256.311	0.813
400	1.330	−230.821	0.879
420	1.439	−203.133	0.946
440	1.545	−173.292	1.016
460	1.650	−141.338	1.087
480	1.752	−107.312	1.159
500	1.853	−71.255	1.233
520	1.952	−33.206	1.307
536.67	2.032	0	1.370
540	2.048	6.794	1.383
560	2.143	48.707	1.459
580	2.236	92.495	1.536
600	2.327	138.121	1.613
620	2.416	185.546	1.691
640	2.503	234.734	1.769
660	2.588	285.650	1.847
680	2.672	338.257	1.926
700	2.754	392.520	2.004
720	2.834	448.404	2.083
740	2.913	505.876	2.162
760	2.990	564.901	2.240
780	3.065	625.447	2.319
800	3.138	687.480	2.398
820	3.210	750.968	2.476
840	3.281	815.880	2.554
860	3.349	882.184	2.632
880	3.417	949.848	2.710
900	3.483	1018.844	2.787
920	3.547	1089.140	2.865
940	3.610	1160.706	2.942
960	3.671	1233.515	3.018
980	3.731	1307.536	3.095
1000	3.789	1382.741	3.171
1020	3.847	1459.103	3.246
1040	3.902	1536.594	3.321
1060	3.957	1615.186	3.396
1080	4.010	1694.854	3.471
1100	4.062	1775.569	3.545
1120	4.112	1857.307	3.618
1140	4.161	1940.041	3.692
1160	4.209	2023.747	3.764
1180	4.256	2108.398	3.837
1200	4.301	2193.972	3.909
1220	4.346	2280.442	3.980
1240	4.389	2367.786	4.051

T (R)	$\bar{c}_p$ (Btu/lbmol·R)	$\bar{h}°(T) - \bar{h}_f°(298)$ (Btu/lbmol)	$\bar{s}°(T)$ (Btu/lbmol·R)
1260	4.431	2455.980	4.122
1280	4.471	2545.000	4.192
1300	4.511	2634.824	4.261
1320	4.549	2725.428	4.331
1340	4.587	2816.791	4.399
1360	4.623	2908.890	4.468
1380	4.658	3001.704	4.535
1400	4.692	3095.211	4.603
1420	4.725	3189.389	4.669
1440	4.757	3284.219	4.736
1460	4.788	3379.679	4.802
1480	4.818	3475.749	4.867
1500	4.847	3572.409	4.932
1520	4.875	3669.639	4.996
1540	4.902	3767.419	5.060
1560	4.929	3865.730	5.123
1580	4.954	3964.552	5.186
1600	4.978	4063.868	5.249
1620	5.001	4163.658	5.311
1640	5.023	4263.903	5.372
1660	5.045	4364.586	5.433
1680	5.065	4465.688	5.494
1700	5.085	4567.191	5.554
1720	5.104	4669.078	5.614
1740	5.122	4771.332	5.673
1760	5.139	4873.934	5.731
1780	5.155	4976.868	5.789
1800	5.170	5080.117	5.847
1820	5.185	5183.666	5.904
1840	5.200	5287.510	5.961
1860	5.214	5391.645	6.017
1880	5.228	5496.068	6.073
1900	5.243	5600.776	6.129
1940	5.270	5811.035	6.238
1980	5.298	6022.396	6.346
2020	5.324	6234.834	6.452
2060	5.350	6448.326	6.557
2100	5.376	6662.849	6.660
2140	5.401	6878.378	6.762
2180	5.425	7094.891	6.862
2220	5.449	7312.367	6.961
2260	5.472	7530.782	7.058
2300	5.495	7750.117	7.154
2340	5.517	7970.349	7.249
2380	5.539	8191.459	7.343
2420	5.560	8413.426	7.436
2460	5.580	8636.230	7.527
2500	5.601	8859.853	7.617
2540	5.620	9084.276	7.706
2580	5.640	9309.480	7.794
2620	5.659	9535.447	7.881

(continued)

Table BE.13 (continued)

T (R)	$\bar{c}_p$ (Btu/lbmol·R)	$\bar{h}°(T) - \bar{h}_f°(298)$ (Btu/lbmol)	$\bar{s}°(T)$ (Btu/lbmol·R)
2660	5.677	9762.159	7.967
2700	5.695	9989.600	8.052
2740	5.713	10217.752	8.136
2780	5.730	10446.599	8.219
2820	5.747	10676.126	8.301
2860	5.763	10906.315	8.382
2900	5.779	11137.153	8.462
2940	5.795	11368.623	8.541
2980	5.810	11600.712	8.619
3020	5.825	11833.405	8.697
3060	5.839	12066.688	8.774
3100	5.854	12300.548	8.850
3140	5.868	12534.972	8.925
3180	5.881	12769.947	8.999
3220	5.894	13005.460	9.073
3260	5.907	13241.500	9.146
3300	5.920	13478.053	9.218
3340	5.933	13715.110	9.289
3380	5.945	13952.658	9.360
3420	5.957	14190.688	9.430
3460	5.968	14429.187	9.499
3500	5.980	14668.147	9.568
3540	5.991	14907.557	9.636
3580	6.002	15147.408	9.703
3620	6.012	15387.690	9.770
3660	6.023	15628.394	9.836
3700	6.033	15869.512	9.902
3740	6.043	16111.034	9.967
3780	6.053	16352.954	10.031
3820	6.063	16595.262	10.095
3860	6.072	16837.952	10.158
3900	6.081	17081.016	10.221
3940	6.090	17324.446	10.283
3980	6.099	17568.237	10.344
4020	6.108	17812.381	10.405
4060	6.117	18056.872	10.466
4100	6.125	18301.704	10.526
4140	6.133	18546.871	10.585
4180	6.142	18792.367	10.644
4220	6.150	19038.188	10.703
4260	6.157	19284.327	10.761
4300	6.165	19530.781	10.818
4340	6.173	19777.543	10.876
4380	6.180	20024.611	10.932
4420	6.188	20271.979	10.988
4460	6.195	20519.643	11.044
4500	6.203	20767.599	11.100
4540	6.210	21015.845	11.155
4580	6.217	21264.376	11.209
4620	6.224	21513.188	11.263
4660	6.231	21762.280	11.317

T (R)	$\bar{c}_p$ (Btu/lbmol·R)	$\bar{h}°(T) - \bar{h}_f°(298)$ (Btu/lbmol)	$\bar{s}°(T)$ (Btu/lbmol·R)
4700	6.238	22011.648	11.370
4740	6.244	22261.288	11.423
4780	6.251	22511.199	11.475
4820	6.258	22761.379	11.528
4860	6.264	23011.824	11.579
4900	6.271	23262.533	11.631
5000	6.287	23890.448	11.758
5100	6.303	24519.973	11.882
5200	6.319	25151.088	12.005
5300	6.335	25783.776	12.125
5400	6.350	26418.028	12.244
5500	6.366	27053.836	12.361
5600	6.381	27691.196	12.475
5700	6.397	28330.109	12.588
5800	6.413	28970.579	12.700
5900	6.428	29612.611	12.810
6000	6.444	30256.213	12.918
6100	6.460	30901.394	13.024
6200	6.476	31548.166	13.130
6300	6.492	32196.540	13.233
6400	6.508	32846.529	13.336
6500	6.524	33498.144	13.437
6600	6.541	34151.399	13.536
6700	6.557	34806.304	13.635
6800	6.574	35462.868	13.732
6900	6.591	36121.101	13.828
7000	6.607	36781.008	13.923
7100	6.624	37442.593	14.017
7200	6.641	38105.858	14.110
7300	6.658	38770.800	14.202
7400	6.674	39437.412	14.292
7500	6.691	40105.685	14.382
7600	6.707	40775.603	14.471
7700	6.724	41447.148	14.558
7800	6.739	42120.293	14.645
7900	6.755	42795.008	14.731
8000	6.770	43471.255	14.816
8100	6.785	44148.992	14.901
8200	6.799	44828.166	14.984
8300	6.812	45508.720	15.066
8400	6.825	46190.587	15.148
8500	6.837	46873.693	15.229
8600	6.848	47557.954	15.309
8700	6.858	48243.278	15.388
8800	6.867	48929.563	15.467
8900	6.875	49616.697	15.544
9000	6.882	50304.557	15.621

Appendix CE
Thermodynamic and
Thermo-Physical Properties of Air

Table CE.1 Approximate Composition, Apparent Molecular Weight, and Gas Constant for Dry Air

Constituent	Mole %
N_2	78.08
O_2	20.95
Ar	0.93
CO_2	0.036
Ne, He, CH_4, others	0.003
m_{air}	28.97 lbm/lbmol
R_{air}	0.06855 Btu/lbm·R = 0.3704 psia·ft³/lbm·R

Table CE.2 Thermodynamic Properties of Air at 1 atm

Temp. (R)	h (Btu/lbm)	u (Btu/lbm)	s° (Btu/lbm·R)	c_p (Btu/lbm·R)	c_v (Btu/lbm·R)	γ (= c_p/c_v)
360	140.16	115.51	0.83166	0.24069	0.17121	1.4058
380	144.97	118.94	0.84467	0.24057	0.17120	1.4052
400	149.78	122.37	0.85701	0.24048	0.17120	1.4047
420	154.59	125.81	0.86874	0.24043	0.17121	1.4042
440	159.40	129.24	0.87993	0.24039	0.17124	1.4038
460	164.20	132.67	0.89061	0.24039	0.17129	1.4034
480	169.01	136.10	0.90084	0.24040	0.17135	1.4030
500	173.82	139.53	0.91066	0.24044	0.17143	1.4026
520	178.63	142.96	0.92009	0.24050	0.17153	1.4021
540	183.44	146.40	0.92917	0.24059	0.17164	1.4017
560	188.25	149.83	0.93792	0.24069	0.17177	1.4012
580	193.07	153.27	0.94637	0.24083	0.17193	1.4007
600	197.89	156.72	0.95453	0.24098	0.17211	1.4002
620	202.71	160.16	0.96244	0.24116	0.17231	1.3996
640	207.53	163.61	0.97010	0.24137	0.17253	1.3990
660	212.36	167.07	0.97753	0.24160	0.17278	1.3983
680	217.20	170.53	0.98475	0.24186	0.17305	1.3976
700	222.04	173.99	0.99176	0.24214	0.17334	1.3969
720	226.88	177.47	0.99859	0.24244	0.17366	1.3961
740	231.74	180.94	1.0052	0.24277	0.17400	1.3953
760	236.59	184.43	1.0117	0.24313	0.17436	1.3944
780	241.46	187.92	1.0180	0.24350	0.17475	1.3935
800	246.34	191.42	1.0242	0.24390	0.17515	1.3925

Temp. (R)	h (Btu/lbm)	u (Btu/lbm)	s° (Btu/lbm·R)	c_p (Btu/lbm·R)	c_v (Btu/lbm·R)	γ (= c_p/c_v)
820	251.22	194.93	1.0302	0.24432	0.17558	1.3915
840	256.11	198.45	1.0361	0.24476	0.17603	1.3905
860	261.01	201.97	1.0419	0.24523	0.17650	1.3894
880	265.92	205.51	1.0475	0.24571	0.17698	1.3883
900	270.84	209.06	1.0531	0.24620	0.17748	1.3872
920	275.77	212.61	1.0585	0.24672	0.17800	1.3860
940	280.71	216.18	1.0638	0.24725	0.17854	1.3848
960	285.66	219.75	1.0690	0.24779	0.17909	1.3836
980	290.62	223.34	1.0741	0.24835	0.17965	1.3824
1000	295.59	226.94	1.0791	0.24892	0.18022	1.3812
1020	300.57	230.55	1.0841	0.24950	0.18081	1.3799
1040	305.57	234.18	1.0889	0.25009	0.18140	1.3787
1060	310.58	237.81	1.0937	0.25069	0.18201	1.3774
1080	315.60	241.46	1.0984	0.25130	0.18262	1.3761
1100	320.63	245.12	1.1030	0.25192	0.18324	1.3748
1120	325.67	248.79	1.1075	0.25254	0.18386	1.3735
1140	330.73	252.47	1.1120	0.25317	0.18449	1.3722
1160	335.80	256.17	1.1164	0.25380	0.18512	1.3710
1180	340.88	259.88	1.1208	0.25443	0.18576	1.3697
1200	345.98	263.60	1.1251	0.25507	0.18640	1.3684
1220	351.09	267.34	1.1293	0.25571	0.18704	1.3671
1240	356.21	271.08	1.1334	0.25635	0.18768	1.3659
1260	361.34	274.84	1.1375	0.25699	0.18832	1.3646
1280	366.49	278.62	1.1416	0.25762	0.18896	1.3634
1300	371.64	282.40	1.1456	0.25826	0.18960	1.3621
1320	376.82	286.20	1.1495	0.25890	0.19024	1.3609
1340	382.00	290.01	1.1534	0.25953	0.19087	1.3597
1360	387.20	293.84	1.1573	0.26016	0.19151	1.3585
1380	392.41	297.67	1.1611	0.26079	0.19214	1.3573
1400	397.63	301.52	1.1649	0.26142	0.19276	1.3562
1420	402.86	305.39	1.1686	0.26204	0.19338	1.3550
1440	408.11	309.26	1.1722	0.26265	0.19400	1.3539
1460	413.37	313.15	1.1759	0.26327	0.19462	1.3528
1500	423.93	320.96	1.1830	0.26448	0.19583	1.3506
1540	434.53	328.81	1.1900	0.26566	0.19701	1.3484
1580	445.18	336.72	1.1968	0.26683	0.19818	1.3464
1620	455.87	344.67	1.2035	0.26796	0.19932	1.3444
1660	466.61	352.66	1.2100	0.26908	0.20043	1.3425
1700	477.40	360.70	1.2164	0.27016	0.20152	1.3406
1740	488.23	368.79	1.2227	0.27122	0.20258	1.3388
1780	499.10	376.91	1.2289	0.27225	0.20361	1.3371
1820	510.01	385.07	1.2350	0.27326	0.20462	1.3354
1860	520.96	393.28	1.2409	0.27423	0.20560	1.3338
1900	531.95	401.52	1.2468	0.27518	0.20655	1.3323
1940	542.97	409.80	1.2525	0.27611	0.20747	1.3308
1980	554.03	418.12	1.2582	0.27700	0.20837	1.3294
2020	565.13	426.47	1.2637	0.27787	0.20924	1.3280
2060	576.26	434.86	1.2692	0.27872	0.21008	1.3267
2100	587.43	443.28	1.2745	0.27954	0.21090	1.3254
2140	598.63	451.73	1.2798	0.28033	0.21170	1.3242
2180	609.85	460.22	1.2850	0.28110	0.21247	1.3230
2220	621.11	468.73	1.2901	0.28185	0.21322	1.3219

(continued)

Table CE.2 (continued)

Temp. (R)	h (Btu/lbm)	u (Btu/lbm)	s° (Btu/lbm·R)	c_p (Btu/lbm·R)	c_v (Btu/lbm·R)	$\gamma\ (= c_p/c_v)$
2260	632.40	477.27	1.2952	0.28258	0.21395	1.3208
2300	643.72	485.85	1.3001	0.28328	0.21465	1.3197
2340	655.06	494.45	1.3050	0.28397	0.21534	1.3187
2380	666.44	503.07	1.3099	0.28463	0.21600	1.3177
2420	677.83	511.73	1.3146	0.28527	0.21664	1.3168
2460	689.26	520.40	1.3193	0.28589	0.21727	1.3159
2500	700.71	529.11	1.3239	0.28650	0.21787	1.3150
2540	712.18	537.83	1.3285	0.28709	0.21846	1.3141
2580	723.67	546.58	1.3329	0.28766	0.21903	1.3133
2620	735.19	555.36	1.3374	0.28821	0.21958	1.3125
2660	746.73	564.15	1.3417	0.28875	0.22012	1.3118
2700	758.29	572.97	1.3461	0.28927	0.22064	1.3110
2740	769.87	581.80	1.3503	0.28978	0.22115	1.3103
2780	781.47	590.66	1.3545	0.29027	0.22164	1.3096
2820	793.09	599.53	1.3587	0.29075	0.22212	1.3090
2860	804.73	608.43	1.3628	0.29121	0.22259	1.3083
2900	816.39	617.34	1.3668	0.29167	0.22304	1.3077
2940	828.06	626.27	1.3708	0.29211	0.22348	1.3071
2980	839.76	635.22	1.3748	0.29254	0.22391	1.3065
3000	845.61	639.70	1.3767	0.29275	0.22412	1.3062
3100	874.94	662.16	1.3863	0.29375	0.22513	1.3048
3200	904.36	684.72	1.3957	0.29470	0.22607	1.3035
3300	933.87	707.38	1.4048	0.29558	0.22696	1.3024
3400	963.47	730.11	1.4136	0.29642	0.22779	1.3013
3500	993.16	752.93	1.4222	0.29720	0.22858	1.3002

Table CE.3   Thermo-Phsyical Properties of Air at 1 atm

Temp. (R)	ρ (lbm/ft^3)	c_p (Btu/lbm·R)	μ (lbm/ft·s)	v (ft^2/s)	k (Btu/h·ft·°F)	α (ft^2/s)	Pr
180	0.22506	0.24869	0.000004776	0.0000212	0.005476	0.0000272	0.78081
200	0.20137	0.24581	0.000005285	0.0000262	0.006094	0.0000342	0.76741
220	0.18236	0.24409	0.000005781	0.0000317	0.006703	0.0000418	0.75792
240	0.16671	0.24299	0.000006266	0.0000376	0.007302	0.0000501	0.75068
260	0.15359	0.24224	0.000006740	0.0000439	0.007891	0.0000589	0.74484
280	0.14242	0.24172	0.000007203	0.0000506	0.008471	0.0000684	0.73995
300	0.13278	0.24134	0.000007656	0.0000577	0.009041	0.0000784	0.73574
320	0.12437	0.24106	0.000008100	0.0000651	0.009602	0.0000890	0.73203
340	0.11698	0.24085	0.000008534	0.0000730	0.010154	0.0001001	0.72871
360	0.11042	0.24069	0.000008960	0.0000811	0.010698	0.0001118	0.72570
380	0.10456	0.24057	0.000009377	0.0000897	0.011233	0.0001241	0.72295
400	0.099299	0.24048	0.000009786	0.0000986	0.011760	0.0001368	0.72041
420	0.094543	0.24043	0.000010188	0.0001078	0.012280	0.0001501	0.71807
440	0.090224	0.24039	0.000010583	0.0001173	0.012793	0.0001638	0.71591
460	0.086283	0.24039	0.000010970	0.0001271	0.013298	0.0001781	0.71389
480	0.082674	0.24040	0.000011351	0.0001373	0.013797	0.0001928	0.71203
500	0.079355	0.24044	0.000011725	0.0001478	0.014289	0.0002080	0.71030
520	0.076294	0.24050	0.000012094	0.0001585	0.014775	0.0002237	0.70870
540	0.073460	0.24059	0.000012457	0.0001696	0.015255	0.0002398	0.70723
560	0.070830	0.24069	0.000012814	0.0001809	0.015729	0.0002563	0.70588
580	0.068382	0.24083	0.000013165	0.0001925	0.016198	0.0002732	0.70465
600	0.066098	0.24098	0.000013512	0.0002044	0.016661	0.0002906	0.70353
620	0.063962	0.24116	0.000013853	0.0002166	0.017120	0.0003083	0.70253
640	0.061960	0.24137	0.000014190	0.0002290	0.017573	0.0003264	0.70163
660	0.060080	0.24160	0.000014522	0.0002417	0.018022	0.0003449	0.70085
680	0.058311	0.24186	0.000014850	0.0002547	0.018466	0.0003637	0.70016
700	0.056643	0.24214	0.000015173	0.0002679	0.018906	0.0003829	0.69958
720	0.055068	0.24244	0.000015493	0.0002813	0.019342	0.0004024	0.69910
740	0.053578	0.24277	0.000015808	0.0002951	0.019773	0.0004223	0.69871
760	0.052167	0.24313	0.000016119	0.0003090	0.020201	0.0004424	0.69841
780	0.050828	0.24350	0.000016427	0.0003232	0.020625	0.0004629	0.69821
800	0.049557	0.24390	0.000016732	0.0003376	0.021045	0.0004837	0.69808
820	0.048347	0.24432	0.000017033	0.0003523	0.021462	0.0005047	0.69804
840	0.047195	0.24476	0.000017330	0.0003672	0.021875	0.0005260	0.69808
860	0.046097	0.24523	0.000017624	0.0003823	0.022285	0.0005476	0.69819
880	0.045049	0.24571	0.000017916	0.0003977	0.022691	0.0005695	0.69837
900	0.044048	0.24620	0.000018204	0.0004133	0.023095	0.0005916	0.69861
920	0.043090	0.24672	0.000018489	0.0004291	0.023496	0.0006139	0.69892
940	0.042173	0.24725	0.000018771	0.0004451	0.023893	0.0006365	0.69929
960	0.041294	0.24779	0.000019051	0.0004614	0.024288	0.0006594	0.69970
980	0.040451	0.24835	0.000019328	0.0004778	0.024680	0.0006824	0.70017
1000	0.039642	0.24892	0.000019602	0.0004945	0.025070	0.0007057	0.70069
1020	0.038864	0.24950	0.000019874	0.0005114	0.025457	0.0007292	0.70124
1040	0.038117	0.25009	0.000020144	0.0005285	0.025841	0.0007530	0.70184
1060	0.037398	0.25069	0.000020411	0.0005458	0.026223	0.0007770	0.70247
1080	0.036705	0.25130	0.000020676	0.0005633	0.026603	0.0008011	0.70313
1100	0.036038	0.25192	0.000020938	0.0005810	0.026980	0.0008255	0.70382
1120	0.035394	0.25254	0.000021199	0.0005989	0.027355	0.0008501	0.70453
1140	0.034773	0.25317	0.000021457	0.0006171	0.027728	0.0008749	0.70527
1160	0.034174	0.25380	0.000021713	0.0006354	0.028099	0.0008999	0.70603
1180	0.033594	0.25443	0.000021967	0.0006539	0.028468	0.0009252	0.70680

(continued)

Table CE.3 (continued)

Temp. (R)	ρ (lbm/ft^3)	c_p (Btu/lbm·R)	μ (lbm/ft·s)	ν (ft^2/s)	k (Btu/h·ft·°F)	α (ft^2/s)	Pr
1200	0.033034	0.25507	0.000022219	0.0006726	0.028835	0.0009506	0.70758
1220	0.032493	0.25571	0.000022470	0.0006915	0.029200	0.0009762	0.70838
1240	0.031969	0.25635	0.000022718	0.0007106	0.029563	0.0010020	0.70919
1260	0.031461	0.25699	0.000022965	0.0007299	0.029924	0.0010281	0.71000
1280	0.030970	0.25762	0.000023210	0.0007494	0.030283	0.0010543	0.71082
1300	0.030493	0.25826	0.000023453	0.0007691	0.030641	0.0010808	0.71165
1320	0.030031	0.25890	0.000023695	0.0007890	0.030997	0.0011074	0.71247
1340	0.029583	0.25953	0.000023935	0.0008091	0.031351	0.0011343	0.71329
1360	0.029148	0.26016	0.000024173	0.0008293	0.031704	0.0011613	0.71411
1380	0.028726	0.26079	0.000024410	0.0008498	0.032055	0.0011886	0.71493
1400	0.028316	0.26142	0.000024645	0.0008704	0.032405	0.0012160	0.71574
1420	0.027917	0.26204	0.000024879	0.0008912	0.032753	0.0012437	0.71655
1440	0.027529	0.26265	0.000025111	0.0009122	0.033100	0.0012716	0.71735
1460	0.027152	0.26327	0.000025342	0.0009334	0.033445	0.0012997	0.71814
1480	0.026785	0.26387	0.000025572	0.0009547	0.033789	0.0013280	0.71893
1500	0.026428	0.26448	0.000025800	0.0009762	0.034132	0.0013565	0.71970
1520	0.026080	0.26507	0.000026027	0.0009980	0.034473	0.0013852	0.72047
1540	0.025742	0.26566	0.000026253	0.0010199	0.034813	0.0014141	0.72122
1560	0.025412	0.26625	0.000026477	0.0010419	0.035152	0.0014432	0.72196
1580	0.025090	0.26683	0.000026701	0.0010642	0.035489	0.0014725	0.72269
1600	0.024777	0.26740	0.000026923	0.0010866	0.035826	0.0015021	0.72341
1620	0.024471	0.26796	0.000027144	0.0011092	0.036161	0.0015318	0.72411
1640	0.024172	0.26852	0.000027363	0.0011320	0.036495	0.0015618	0.72480
1660	0.023881	0.26908	0.000027582	0.0011550	0.036828	0.0015920	0.72548
1680	0.023597	0.26962	0.000027799	0.0011781	0.037160	0.0016224	0.72614
1700	0.023319	0.27016	0.000028016	0.0012014	0.037491	0.0016530	0.72679
1720	0.023048	0.27070	0.000028231	0.0012249	0.037820	0.0016838	0.72742
1740	0.022783	0.27122	0.000028446	0.0012485	0.038149	0.0017149	0.72804
1760	0.022525	0.27174	0.000028659	0.0012723	0.038477	0.0017462	0.72865
1780	0.022272	0.27225	0.000028871	0.0012963	0.038803	0.0017776	0.72924
1800	0.022024	0.27276	0.000029083	0.0013205	0.039129	0.0018093	0.72982
1900	0.020865	0.27518	0.000030126	0.0014438	0.040745	0.0019712	0.73248
2000	0.019822	0.27744	0.000031148	0.0015714	0.042339	0.0021385	0.73478
2100	0.018878	0.27954	0.000032151	0.0017030	0.043915	0.0023116	0.73675
2200	0.018021	0.28148	0.000033136	0.0018388	0.045474	0.0024903	0.73839
2300	0.017237	0.28328	0.000034105	0.0019786	0.047018	0.0026747	0.73975
2400	0.016519	0.28495	0.000035060	0.0021224	0.048547	0.0028649	0.74083
2500	0.015859	0.28650	0.000036001	0.0022702	0.050065	0.0030608	0.74168
2600	0.015249	0.28794	0.000036931	0.0024219	0.051570	0.0032626	0.74231
2700	0.014684	0.28927	0.000037849	0.0025776	0.053066	0.0034703	0.74275
2800	0.014160	0.29051	0.000038757	0.0027371	0.054552	0.0036838	0.74303
2900	0.013672	0.29167	0.000039656	0.0029006	0.056030	0.0039031	0.74315
3000	0.013216	0.29275	0.000040546	0.0030679	0.057500	0.0041283	0.74314
3200	0.012390	0.29470	0.000042302	0.0034142	0.060419	0.0045964	0.74280
3400	0.011661	0.29642	0.000044033	0.0037759	0.063316	0.0050881	0.74211
3600	0.011014	0.29794	0.000045740	0.0041530	0.066193	0.0056034	0.74116

Appendix DE
Thermodynamic Properties of H$_2$O

Table DE.1 Saturated Properties of Water and Steam: Temperature Increments

Temp., °F	Press., P_{sat}, psia	Specific Volume, ft³/lbm		Internal Energy, Btu/lbm			Enthalpy, Btu/lbm			Entropy, Btu/lbm·R		
		Sat. Liquid, v_f	Sat. Vapor, v_g	Sat. Liquid, u_f	Evap., u_{fg}	Sat. Vapor, u_g	Sat. Liquid, h_f	Evap., h_{fg}	Sat. Vapor, h_g	Sat. Liquid, s_f	Evap., s_{fg}	Sat. Vapor, s_g
32.018	0.088713	0.016022	3299.7	0	1021.7	1021.7	0.00026319	1075.9	1075.9	0	2.1882	2.1882
34	0.096068	0.016021	3059.2	1.9981	1020.4	1022.4	1.9984	1074.8	1076.8	0.0040557	2.1771	2.1812
36	0.10403	0.016020	2836.4	4.0127	1019.0	1023.0	4.0130	1073.7	1077.7	0.0081282	2.1661	2.1742
38	0.11258	0.016020	2631.6	6.0257	1017.7	1023.7	6.0261	1072.5	1078.5	0.012181	2.1551	2.1673
40	0.12173	0.016020	2443.3	8.0373	1016.3	1024.3	8.0377	1071.4	1079.4	0.016215	2.1442	2.1604
42	0.13155	0.016020	2270.1	10.0480	1015.0	1025.0	10.048	1070.3	1080.3	0.020230	2.1334	2.1536
44	0.14205	0.016021	2110.5	12.0570	1013.6	1025.7	12.057	1069.1	1081.2	0.024227	2.1227	2.1469
46	0.15329	0.016021	1963.5	14.0650	1012.2	1026.3	14.065	1067.9	1082.0	0.028206	2.1120	2.1402
48	0.16530	0.016023	1827.9	16.0710	1010.9	1027.0	16.072	1066.8	1082.9	0.032167	2.1014	2.1336
50	0.17814	0.016024	1702.8	18.0770	1009.5	1027.6	18.078	1065.7	1083.8	0.036110	2.0910	2.1271
52	0.19184	0.016026	1587.3	20.0830	1008.2	1028.3	20.083	1064.6	1084.7	0.040037	2.0806	2.1206
54	0.20647	0.016028	1480.5	22.0870	1006.8	1028.9	22.087	1063.4	1085.5	0.043946	2.0703	2.1142
56	0.22207	0.016030	1381.9	24.0900	1005.5	1029.6	24.091	1062.3	1086.4	0.047839	2.0601	2.1079
58	0.23869	0.016032	1290.5	26.0930	1004.1	1030.2	26.094	1061.2	1087.3	0.051716	2.0499	2.1016
60	0.25640	0.016035	1206.0	28.0960	1002.8	1030.9	28.096	1060.1	1088.2	0.055576	2.0398	2.0954
62	0.27525	0.016038	1127.7	30.0970	1001.4	1031.5	30.098	1058.9	1089.0	0.059421	2.0299	2.0893
64	0.29530	0.016041	1055.0	32.0990	1000.1	1032.2	32.099	1057.8	1089.9	0.063250	2.0200	2.0832
66	0.31663	0.016044	987.69	34.0990	998.80	1032.9	34.100	1056.7	1090.8	0.067063	2.0101	2.0772
68	0.33929	0.016048	925.17	36.1000	997.40	1033.5	36.101	1055.5	1091.6	0.070861	2.0003	2.0712
70	0.36336	0.016052	867.11	38.1000	996.10	1034.2	38.101	1054.4	1092.5	0.074644	1.9907	2.0653
72	0.38891	0.016056	813.16	40.099	994.70	1034.8	40.101	1053.3	1093.4	0.078412	1.9811	2.0595
74	0.41601	0.016060	762.99	42.099	993.40	1035.5	42.100	1052.1	1094.2	0.082166	1.9715	2.0537
76	0.44475	0.016064	716.31	44.098	992.00	1036.1	44.099	1051.0	1095.1	0.085905	1.9620	2.0479
78	0.47521	0.016069	672.86	46.097	990.70	1036.8	46.098	1049.9	1096.0	0.089629	1.9527	2.0423
80	0.50747	0.016074	632.38	48.095	989.31	1037.4	48.097	1048.7	1096.8	0.093340	1.9433	2.0366
82	0.54162	0.016079	594.66	50.094	988.01	1038.1	50.095	1047.6	1097.7	0.097036	1.9341	2.0311
84	0.57776	0.016084	559.48	52.092	986.61	1038.7	52.094	1046.5	1098.6	0.10072	1.9248	2.0255
86	0.61597	0.016089	526.66	54.09	985.31	1039.4	54.092	1045.3	1099.4	0.10439	1.9157	2.0201
88	0.65637	0.016095	496.02	56.088	983.91	1040.0	56.090	1044.2	1100.3	0.10804	1.9067	2.0147
90	0.69904	0.016100	467.40	58.088	982.51	1040.6	58.088	1043.0	1101.1	0.11168	1.8976	2.0093
92	0.74410	0.016106	440.65	60.084	981.22	1041.3	60.086	1041.9	1102.0	0.11531	1.8887	2.0040
94	0.79167	0.016112	415.64	62.082	979.82	1041.9	62.084	1040.8	1102.9	0.11893	1.8798	1.9987
96	0.84184	0.016118	392.25	64.08	978.52	1042.6	64.082	1039.6	1103.7	0.12253	1.8710	1.9935
98	0.89475	0.016125	370.34	66.078	977.12	1043.2	66.080	1038.5	1104.6	0.12612	1.8622	1.9883
100	0.95051	0.016131	349.83	68.076	975.82	1043.9	68.078	1037.3	1105.4	0.12969	1.8535	1.9832
110	1.2767	0.016166	264.97	78.065	969.04	1047.1	78.069	1031.6	1109.7	0.14738	1.8109	1.9583
120	1.6950	0.016205	202.95	88.057	962.24	1050.3	88.062	1025.9	1114.0	0.16477	1.7698	1.9346
130	2.2259	0.016247	157.09	98.052	955.35	1053.4	98.059	1020.1	1118.2	0.18187	1.7299	1.9118

140	2.8930	0.016293	122.82	108.05	948.55	1056.6	108.06	1014.2	1122.3	0.19868	1.6914	1.8901
150	3.7232	0.016342	96.930	118.06	941.64	1059.7	118.07	1008.4	1126.5	0.21523	1.6541	1.8693
160	4.7472	0.016394	77.184	128.07	934.63	1062.7	128.08	1002.5	1130.6	0.23152	1.6178	1.8493
170	5.9998	0.016449	61.980	138.09	927.71	1065.8	138.11	996.49	1134.6	0.24756	1.5826	1.8302
180	7.5195	0.016508	50.169	148.12	920.68	1068.8	148.14	990.46	1138.6	0.26336	1.5484	1.8118
190	9.3496	0.016569	40.916	158.16	913.64	1071.8	158.18	984.42	1142.6	0.27893	1.5153	1.7942
200	11.538	0.016633	33.609	168.21	906.49	1074.7	168.24	978.26	1146.5	0.29429	1.4829	1.7772
210	14.136	0.016701	27.794	178.27	899.23	1077.5	178.31	971.99	1150.3	0.30943	1.4515	1.7609
220	17.201	0.016771	23.133	188.35	892.05	1080.4	188.40	965.70	1154.1	0.32436	1.4207	1.7451
230	20.795	0.016845	19.371	198.44	884.66	1083.1	198.51	959.19	1157.7	0.33911	1.3908	1.7299
240	24.986	0.016921	16.314	208.55	877.25	1085.8	208.63	952.67	1161.3	0.35366	1.3616	1.7153
250	29.844	0.017001	13.815	218.68	869.82	1088.5	218.78	946.02	1164.8	0.36804	1.3331	1.7011
260	35.447	0.017084	11.759	228.83	862.17	1091.0	228.95	939.25	1168.2	0.38225	1.3052	1.6874
270	41.878	0.017170	10.058	239.01	854.49	1093.5	239.14	932.36	1171.5	0.39629	1.2778	1.6741
280	49.222	0.017259	8.6431	249.21	846.69	1095.9	249.37	925.33	1174.7	0.41017	1.2510	1.6612
290	57.574	0.017352	7.4600	259.44	838.86	1098.3	259.62	918.18	1177.8	0.42391	1.2248	1.6487
300	67.029	0.017448	6.4658	269.69	830.81	1100.5	269.91	910.79	1180.7	0.43750	1.1990	1.6365
310	77.691	0.017548	5.6263	279.98	822.62	1102.6	280.23	903.37	1183.6	0.45096	1.1736	1.6246
320	89.667	0.017652	4.9142	290.30	814.40	1104.7	290.60	895.70	1186.3	0.46428	1.1488	1.6131
330	103.07	0.017760	4.3075	300.66	805.94	1106.6	301.00	887.80	1188.8	0.47749	1.1243	1.6018
340	118.02	0.017872	3.7884	311.06	797.44	1108.5	311.45	879.85	1191.3	0.49058	1.1002	1.5908
350	134.63	0.017987	3.3425	321.50	788.70	1110.2	321.95	871.55	1193.5	0.50355	1.0765	1.5800
360	153.03	0.018108	2.9580	331.99	779.81	1111.8	332.50	863.10	1195.6	0.51643	1.0530	1.5694
370	173.36	0.018233	2.6252	342.52	770.78	1113.3	343.11	854.39	1197.5	0.52921	1.0299	1.5591
380	195.74	0.018363	2.3361	353.11	761.49	1114.6	353.77	845.53	1199.3	0.54190	1.0070	1.5489
390	220.33	0.018498	2.0841	363.75	752.05	1115.8	364.50	836.40	1200.9	0.55450	0.98430	1.5388
400	247.26	0.018639	1.8638	374.45	742.45	1116.9	375.30	826.90	1202.2	0.56703	0.96187	1.5289
410	276.68	0.018785	1.6706	385.21	732.59	1117.8	386.17	817.23	1203.4	0.57948	0.93972	1.5192
420	308.76	0.018938	1.5006	396.04	722.56	1118.6	397.12	807.28	1204.4	0.59187	0.91773	1.5096
430	343.64	0.019097	1.3505	406.94	712.26	1119.2	408.15	796.95	1205.1	0.60420	0.89580	1.5000
440	381.48	0.019263	1.2177	417.91	701.69	1119.6	419.27	786.33	1205.6	0.61648	0.87402	1.4905
450	422.46	0.019437	1.0999	428.97	690.93	1119.9	430.49	775.41	1205.9	0.62872	0.85238	1.4811
460	466.75	0.019619	0.99506	440.11	679.79	1119.9	441.81	764.09	1205.9	0.64092	0.83088	1.4718
470	514.52	0.019810	0.90154	451.35	668.45	1119.8	453.24	752.46	1205.7	0.65309	0.80931	1.4624
480	565.95	0.020010	0.81791	462.69	656.71	1119.4	464.78	740.32	1205.1	0.66524	0.78786	1.4531
490	621.23	0.020220	0.74293	474.13	644.67	1118.8	476.46	727.84	1204.3	0.67738	0.76642	1.4438
500	680.55	0.020441	0.67555	485.69	632.31	1118.0	488.27	714.83	1203.1	0.68952	0.74488	1.4344
520	812.10	0.020923	0.56007	509.21	606.29	1115.5	512.35	687.45	1199.8	0.71384	0.70166	1.4155
540	962.24	0.021464	0.46553	533.32	578.58	1111.9	537.14	657.66	1194.8	0.73829	0.65791	1.3962
560	1132.7	0.022080	0.38742	558.14	548.66	1106.8	562.77	625.23	1188.0	0.76299	0.61321	1.3762
580	1325.5	0.022791	0.32227	583.85	516.05	1099.9	589.44	589.56	1179.0	0.78812	0.56708	1.3552
600	1542.5	0.023631	0.26739	610.67	480.33	1091.0	617.42	549.98	1167.4	0.81388	0.51902	1.3329
620	1786.2	0.024647	0.22062	638.96	440.34	1079.3	647.11	505.19	1152.3	0.84062	0.46788	1.3085
640	2059.2	0.025930	0.18017	669.30	394.60	1063.9	679.19	453.41	1132.6	0.86888	0.41232	1.2812
660	2364.9	0.027661	0.14439	702.84	339.96	1042.8	714.96	391.14	1106.1	0.89974	0.34936	1.2491
680	2707.3	0.030361	0.11132	742.67	269.13	1011.8	757.89	309.71	1067.6	0.93611	0.27169	1.2078
700	3093.0	0.036652	0.074795	802.01	146.81	948.82	823.00	168.66	991.66	0.99065	0.14545	1.1361

Table DE.2 Saturated Properties of Water and Steam: Pressure Increments

Press., P_{sat}, psia	Temp., °F	Specific Volume, ft³/lbm Sat. Liquid, v_f	Sat. Vapor, v_g	Internal Energy, Btu/lbm Sat. Liquid, u_f	Evap., u_{fg}	Sat. Vapor, u_g	Enthalpy, Btu/lbm Sat. Liquid, h_f	Evap., h_{fg}	Sat. Vapor, h_g	Entropy, Btu/lbm·R Sat. Liquid, s_f	Evap., s_{fg}	Sat. Vapor, s_g
1	101.69	0.016137	333.50	69.77	974.63	1044.4	69.769	1036.4	1106.2	0.13271	1.8462	1.9789
2	126.03	0.016230	173.71	94.08	958.12	1052.2	94.085	1022.4	1116.5	0.17511	1.7456	1.9207
3	141.42	0.016300	118.70	109.47	947.53	1057.0	109.48	1013.4	1122.9	0.20104	1.6861	1.8871
4	152.91	0.016357	90.626	120.97	939.63	1060.6	120.98	1006.7	1127.7	0.22000	1.6434	1.8634
5	162.18	0.016406	73.521	130.26	933.14	1063.4	130.27	1001.2	1131.5	0.23504	1.6101	1.8451
6	170.00	0.016449	61.977	138.09	927.71	1065.8	138.11	996.49	1134.6	0.24756	1.5826	1.8302
7	176.79	0.016489	53.647	144.90	922.90	1067.8	144.92	992.48	1137.4	0.25832	1.5593	1.8176
8	182.81	0.016525	47.343	150.93	918.67	1069.6	150.96	988.84	1139.8	0.26776	1.5390	1.8068
9	188.22	0.016558	42.402	156.37	914.83	1071.2	156.40	985.50	1141.9	0.27619	1.5211	1.7973
10	193.16	0.016589	38.421	161.33	911.37	1072.7	161.36	982.44	1143.8	0.28381	1.5049	1.7887
11	197.70	0.016618	35.143	165.89	908.11	1074.0	165.93	979.67	1145.6	0.29078	1.4903	1.7811
12	201.91	0.016646	32.396	170.13	905.07	1075.2	170.16	977.04	1147.2	0.29719	1.4768	1.7740
13	205.83	0.016672	30.059	174.08	902.32	1076.4	174.12	974.58	1148.7	0.30315	1.4645	1.7676
14	209.52	0.016697	28.046	177.78	899.62	1077.4	177.83	972.27	1150.1	0.30870	1.4529	1.7616
14.696	211.95	0.016714	26.802	180.24	897.86	1078.1	180.28	970.72	1151.0	0.31236	1.4454	1.7578
15	212.99	0.016722	26.294	181.28	897.12	1078.4	181.33	970.07	1151.4	0.31391	1.4422	1.7561
16	216.27	0.016745	24.753	184.59	894.71	1079.3	184.64	968.06	1152.7	0.31882	1.4321	1.7509
17	219.39	0.016767	23.389	187.73	892.47	1080.2	187.79	966.01	1153.8	0.32346	1.4226	1.7461
18	222.36	0.016788	22.171	190.73	890.27	1081.0	190.79	964.11	1154.9	0.32787	1.4136	1.7415
19	225.20	0.016809	21.078	193.59	888.21	1081.8	193.65	962.35	1156.0	0.33206	1.4051	1.7372
20	227.92	0.016829	20.091	196.34	886.26	1082.6	196.40	960.60	1157.0	0.33605	1.3971	1.7331
25	240.03	0.016922	16.305	208.58	877.22	1085.8	208.66	952.64	1161.3	0.35371	1.3615	1.7152
30	250.30	0.017003	13.747	218.98	869.52	1088.5	219.08	945.82	1164.9	0.36847	1.3322	1.7007
35	259.25	0.017078	11.899	228.07	862.73	1090.8	228.18	939.82	1168.0	0.38119	1.3072	1.6884
40	267.22	0.017146	10.500	236.17	856.63	1092.8	236.30	934.30	1170.6	0.39240	1.2853	1.6777
45	274.41	0.017209	9.4019	243.51	851.09	1094.6	243.65	929.25	1172.9	0.40243	1.2660	1.6684
50	280.99	0.017268	8.5168	250.22	845.98	1096.2	250.38	924.62	1175.0	0.41153	1.2484	1.6599
55	287.05	0.017324	7.7876	256.42	841.18	1097.6	256.59	920.31	1176.9	0.41987	1.2324	1.6523
60	292.68	0.017378	7.1761	262.19	836.71	1098.9	262.38	916.22	1178.6	0.42757	1.2178	1.6454
65	297.95	0.017428	6.6556	267.59	832.41	1100.0	267.80	912.40	1180.2	0.43473	1.2043	1.6390
70	302.91	0.017477	6.2071	272.68	828.42	1101.1	272.91	908.69	1181.6	0.44143	1.1916	1.6330
75	307.58	0.017524	5.8165	277.49	824.61	1102.1	277.74	905.16	1182.9	0.44772	1.1798	1.6275
80	312.02	0.017569	5.4730	282.06	821.04	1103.1	282.32	901.78	1184.1	0.45366	1.1686	1.6223
85	316.24	0.017613	5.1687	286.41	817.49	1103.9	286.69	898.61	1185.3	0.45928	1.1581	1.6174
90	320.26	0.017655	4.8970	290.57	814.13	1104.7	290.87	895.43	1186.3	0.46463	1.1482	1.6128
95	324.11	0.017696	4.6530	294.56	810.94	1105.5	294.87	892.43	1187.3	0.46973	1.1387	1.6084
100	327.81	0.017736	4.4326	298.38	807.82	1106.2	298.71	889.59	1188.3	0.47460	1.1297	1.6043

110	334.77	0.017813	4.0498	305.62	801.88	1107.5	305.98	884.02	1190.0	0.48375	1.1128	1.5965
120	341.25	0.017886	3.7288	312.36	796.34	1108.7	312.76	878.74	1191.5	0.49221	1.0972	1.5894
130	347.32	0.017956	3.4556	318.70	791.00	1109.7	319.13	873.77	1192.9	0.50008	1.0828	1.5829
140	353.03	0.018023	3.2201	324.67	786.03	1110.7	325.14	869.06	1194.2	0.50746	1.0693	1.5768
150	358.42	0.018089	3.0150	330.33	781.17	1111.5	330.83	864.47	1195.3	0.51440	1.0567	1.5711
160	363.54	0.018152	2.8347	335.71	776.59	1112.3	336.25	860.05	1196.3	0.52097	1.0447	1.5657
170	368.41	0.018213	2.6748	340.85	772.15	1113.0	341.42	855.78	1197.2	0.52719	1.0335	1.5607
180	373.07	0.018272	2.5321	345.76	767.94	1113.7	346.37	851.73	1198.1	0.53311	1.0228	1.5559
190	377.52	0.018330	2.4040	350.48	763.82	1114.3	351.13	847.77	1198.9	0.53876	1.0126	1.5514
200	381.80	0.018387	2.2882	355.02	759.78	1114.8	355.70	843.90	1199.6	0.54417	1.0028	1.5470
250	400.97	0.018653	1.8440	375.49	741.51	1117.0	376.35	826.05	1202.4	0.56824	0.95976	1.5280
300	417.35	0.018897	1.5435	393.16	725.24	1118.4	394.21	809.89	1204.1	0.58859	0.92351	1.5121
350	431.74	0.019125	1.3263	408.84	710.46	1119.3	410.08	795.12	1205.2	0.60634	0.89206	1.4984
400	444.62	0.019342	1.1616	423.00	696.80	1119.8	424.44	781.36	1205.8	0.62213	0.86407	1.4862
450	456.31	0.019551	1.0324	435.99	683.91	1119.9	437.62	768.38	1206.0	0.63642	0.83878	1.4752
500	467.04	0.019752	0.92815	448.01	671.79	1119.8	449.84	755.96	1205.8	0.64949	0.81571	1.4652
550	476.98	0.019948	0.84223	459.25	660.35	1119.6	461.28	744.02	1205.3	0.66157	0.79433	1.4559
600	486.24	0.020140	0.77015	469.82	649.28	1119.1	472.06	732.54	1204.6	0.67282	0.77448	1.4473
650	494.94	0.020328	0.70878	479.82	638.58	1118.4	482.27	721.53	1203.8	0.68337	0.75583	1.4392
700	503.13	0.020513	0.65586	489.34	628.36	1117.7	492.00	710.70	1202.7	0.69332	0.73818	1.4315
750	510.89	0.020697	0.60974	498.43	618.37	1116.8	501.31	700.19	1201.5	0.70275	0.72145	1.4242
800	518.27	0.020879	0.56917	507.15	608.65	1115.8	510.24	689.86	1200.1	0.71173	0.70537	1.4171
850	525.30	0.021060	0.53319	515.53	599.17	1114.7	518.85	679.75	1198.6	0.72030	0.69010	1.4104
900	532.02	0.021240	0.50106	523.62	589.88	1113.5	527.16	669.84	1197.0	0.72851	0.67539	1.4039
950	538.46	0.021420	0.47217	531.44	580.76	1112.2	535.21	659.99	1195.2	0.73640	0.66130	1.3977
1000	544.65	0.021600	0.44605	539.02	571.78	1110.8	543.02	650.38	1193.4	0.74400	0.64760	1.3916
1100	556.35	0.021961	0.40062	553.55	554.25	1107.8	558.02	631.38	1189.4	0.75846	0.62144	1.3799
1200	567.26	0.022326	0.36243	567.36	537.14	1104.5	572.32	612.68	1185.0	0.77205	0.59665	1.3687
1300	577.49	0.022696	0.32983	580.57	520.33	1100.9	586.03	594.27	1180.3	0.78493	0.57297	1.3579
1400	587.14	0.023074	0.30163	593.28	503.72	1097.0	599.26	575.94	1175.2	0.79722	0.55018	1.3474
1500	596.26	0.023462	0.27697	605.56	487.34	1092.9	612.08	557.72	1169.8	0.80900	0.52820	1.3372
1600	604.93	0.023862	0.25518	617.49	470.91	1088.4	624.56	539.44	1164.0	0.82036	0.50674	1.3271
1700	613.18	0.024277	0.23577	629.13	454.57	1083.7	636.77	521.13	1157.9	0.83137	0.48573	1.3171
1800	621.07	0.024708	0.21832	640.52	438.08	1078.6	648.76	502.64	1151.4	0.84208	0.46512	1.3072
1900	628.61	0.025160	0.20252	651.73	421.47	1073.2	660.58	483.92	1144.5	0.85255	0.44465	1.2972
2000	635.85	0.025635	0.18813	662.79	404.71	1067.5	672.28	464.92	1137.2	0.86284	0.42436	1.2872
2200	649.49	0.026677	0.16271	684.71	370.09	1054.8	695.57	425.53	1121.1	0.88309	0.38361	1.2667
2400	662.16	0.027890	0.14073	706.75	333.35	1040.1	719.15	383.45	1102.6	0.90332	0.34188	1.2452
2600	673.98	0.029380	0.12116	729.65	293.05	1022.7	743.79	337.21	1081.0	0.92423	0.29747	1.2217
2800	685.04	0.031363	0.10300	754.58	246.42	1001.0	770.85	283.55	1054.4	0.94699	0.24771	1.1947
3000	695.41	0.034339	0.084657	783.94	186.64	970.58	803.02	214.58	1017.6	0.97391	0.18579	1.1597
3200	705.10	0.048459	0.051118	861.93	10.78	872.71	890.64	12.360	903.00	1.0481	0.010600	1.0587

Table DE.3   Superheated Vapor (Steam)

Table DE.3A $P = 1$ psia ($T_{sat} = 101.69°F$)

T, °F	ρ, lbm/ft³	v, ft³/lbm	u, Btu/lbm	h, Btu/lbm	s, Btu/lbm·R
Sat.	0.0029985	333.50	1044.4	1106.2	1.9789
200	0.0025476	392.52	1078.2	1150.9	2.0523
240	0.0024013	416.44	1092.0	1169.1	2.0791
280	0.0022710	440.33	1105.8	1187.3	2.1045
320	0.0021542	464.20	1119.7	1205.6	2.1286
360	0.0020489	488.07	1133.7	1224.1	2.1517
400	0.0019534	511.92	1147.8	1242.6	2.1737
440	0.0018665	535.77	1162.1	1261.3	2.1949
480	0.0017869	559.62	1176.4	1280.0	2.2153
520	0.0017139	583.46	1190.9	1298.9	2.2350
560	0.0016466	607.30	1205.5	1317.9	2.2541
600	0.0015844	631.14	1220.2	1337.1	2.2725
640	0.0015268	654.97	1235.1	1356.4	2.2903
680	0.0014732	678.81	1250.1	1375.8	2.3077
720	0.0014232	702.64	1265.2	1395.3	2.3246
760	0.0013765	726.48	1280.5	1415.0	2.3410
800	0.0013328	750.31	1295.9	1434.9	2.3570
900	0.0012347	809.89	1335.1	1485.1	2.3953
1000	0.0011501	869.47	1375.2	1536.2	2.4316
1100	0.0010764	929.04	1416.1	1588.2	2.4660
1200	0.0010115	988.62	1458.1	1641.1	2.4989
1300	0.00095402	1048.2	1500.9	1695.0	2.5305
1400	0.00090272	1107.8	1544.8	1749.9	2.5608

Table DE.3B $P = 5$ psia ($T_{sat} = 162.18°F$)

T, °F	ρ, lbm/ft³	v, ft³/lbm	u, Btu/lbm	h, Btu/lbm	s, Btu/lbm·R
Sat.	0.013602	73.521	1063.4	1131.5	1.8451
200	0.012795	78.153	1076.9	1149.3	1.8729
240	0.012047	83.009	1091.0	1167.9	1.9002
280	0.011385	87.838	1105.0	1186.4	1.9259
320	0.010793	92.650	1119.1	1204.9	1.9503
360	0.010262	97.452	1133.2	1223.5	1.9735
400	0.0097804	102.25	1147.4	1242.1	1.9957
440	0.0093429	107.03	1161.7	1260.8	2.0170
480	0.0089432	111.82	1176.1	1279.6	2.0375
520	0.0085765	116.60	1190.6	1298.6	2.0572
560	0.0082389	121.38	1205.3	1317.6	2.0763
600	0.0079269	126.15	1220.0	1336.8	2.0948
640	0.0076378	130.93	1234.9	1356.1	2.1127
680	0.0073692	135.70	1249.9	1375.6	2.1300
720	0.0071188	140.47	1265.1	1395.2	2.1469
760	0.0068849	145.25	1280.4	1414.9	2.1633
800	0.0066659	150.02	1295.8	1434.7	2.1793
900	0.0061751	161.94	1335.0	1484.9	2.2177
1000	0.0057516	173.86	1375.1	1536.1	2.2540
1100	0.0053826	185.78	1416.1	1588.1	2.2885
1200	0.0050581	197.70	1458.0	1641.1	2.3214
1300	0.0047705	209.62	1500.9	1695.0	2.3529
1400	0.0045139	221.54	1544.7	1749.8	2.3832

Table DE.3C $P = 10$ psia ($T_{sat} = 193.16°F$)

T, °F	ρ, lbm/ft³	v, ft³/lbm	u, Btu/lbm	h, Btu/lbm	s, Btu/lbm·R
Sat.	0.026027	38.421	1072.7	1143.8	1.7887
200	0.025741	38.849	1075.2	1147.2	1.7938
240	0.024198	41.326	1089.8	1166.3	1.8220
280	0.022844	43.774	1104.1	1185.2	1.8482
320	0.021643	46.205	1118.4	1203.9	1.8729
360	0.020566	48.624	1132.6	1222.7	1.8964
400	0.019594	51.035	1146.9	1241.4	1.9187
440	0.018712	53.440	1161.3	1260.2	1.9401
480	0.017908	55.842	1175.8	1279.2	1.9607
520	0.017170	58.240	1190.3	1298.2	1.9805
560	0.016492	60.635	1205.0	1317.3	1.9996
600	0.015866	63.029	1219.8	1336.5	2.0181
640	0.015286	65.421	1234.7	1355.8	2.0360
680	0.014747	67.812	1249.7	1375.3	2.0534
720	0.014245	70.202	1264.9	1394.9	2.0703
760	0.013776	72.591	1280.2	1414.6	2.0867
800	0.013337	74.979	1295.7	1434.5	2.1028
900	0.012354	80.948	1334.9	1484.8	2.1412
1000	0.011506	86.913	1375.0	1535.9	2.1774
1100	0.010767	92.877	1416.0	1588.0	2.2119
1200	0.010117	98.839	1457.9	1641.0	2.2449
1300	0.0095419	104.80	1500.8	1694.9	2.2764
1400	0.0090284	110.76	1544.6	1749.7	2.3067

Table DE.3D $P = 14.696$ psia ($T_{sat} = 211.95°F$)

T, °F	ρ, lbm/ft³	v, ft³/lbm	u, Btu/lbm	h, Btu/lbm	s, Btu/lbm·R
Sat.	0.037311	26.802	1078.1	1151.0	1.7578
240	0.035710	28.004	1088.6	1164.8	1.7778
280	0.033679	29.692	1103.2	1184.0	1.8045
320	0.031885	31.363	1117.7	1203.0	1.8295
360	0.030284	33.021	1132.1	1221.9	1.8532
400	0.028843	34.671	1146.5	1240.8	1.8757
440	0.027537	36.315	1160.9	1259.7	1.8972
500	0.025791	38.774	1182.7	1288.2	1.9279
600	0.023333	42.858	1219.5	1336.2	1.9754
700	0.021307	46.933	1257.1	1384.8	2.0193
800	0.019607	51.002	1295.5	1434.3	2.0602
900	0.018160	55.067	1334.8	1484.6	2.0986
1000	0.016912	59.129	1374.9	1535.8	2.1349
1100	0.015826	63.189	1415.9	1587.9	2.1694
1200	0.014870	67.248	1457.9	1640.9	2.2024
1300	0.014024	71.306	1500.8	1694.8	2.2339
1400	0.013269	75.363	1544.6	1749.7	2.2642
1500	0.012591	79.420	1589.4	1805.5	2.2935
1600	0.011980	83.476	1635.0	1862.2	2.3217

Table DE.3E $P = 20$ psia ($T_{sat} = 227.92°F$)

T, °F	ρ, lbm/ft³	v, ft³/lbm	u, Btu/lbm	h, Btu/lbm	s, Btu/lbm·R
Sat.	0.049774	20.091	1082.6	1157.0	1.7331
240	0.048834	20.478	1087.2	1163.0	1.7418
280	0.046001	21.739	1102.2	1182.7	1.7691
320	0.043516	22.980	1116.9	1202.0	1.7945
360	0.041307	24.209	1131.4	1221.1	1.8184
400	0.039325	25.429	1145.9	1240.1	1.8411
440	0.037532	26.644	1160.4	1259.1	1.8627
500	0.03514	28.458	1182.3	1287.7	1.8935
600	0.031779	31.467	1219.3	1335.8	1.9412
700	0.029013	34.467	1256.9	1384.6	1.9851
800	0.026695	37.461	1295.4	1434.1	2.0261
900	0.024721	40.451	1334.6	1484.4	2.0645
1000	0.023021	43.438	1374.8	1535.6	2.1008
1100	0.021541	46.423	1415.8	1587.7	2.1354
1200	0.02024	49.407	1457.8	1640.8	2.1683
1300	0.019088	52.390	1500.7	1694.7	2.1999
1400	0.01806	55.372	1544.5	1749.6	2.2302
1500	0.017137	58.354	1589.3	1805.4	2.2594
1600	0.016304	61.335	1635.0	1862.1	2.2877

Table DE.3F $P = 40$ psia ($T_{sat} = 267.22°F$)

T, °F	ρ, lbm/ft³	v, ft³/lbm	u, Btu/lbm	h, Btu/lbm	s, Btu/lbm·R
Sat.	0.095239	10.500	1092.8	1170.6	1.6777
280	0.093347	10.713	1098.0	1177.4	1.6870
320	0.088005	11.363	1113.7	1197.8	1.7139
360	0.083343	11.999	1128.9	1217.7	1.7388
400	0.079211	12.625	1143.8	1237.3	1.7622
440	0.075505	13.244	1158.7	1256.8	1.7843
500	0.070596	14.165	1181.0	1285.9	1.8156
600	0.06375	15.686	1218.3	1334.5	1.8637
700	0.05815	17.197	1256.2	1383.5	1.9080
800	0.053472	18.702	1294.7	1433.3	1.9491
900	0.0495	20.202	1334.1	1483.8	1.9877
1000	0.046083	21.700	1374.3	1535.1	2.0241
1100	0.04311	23.196	1415.5	1587.3	2.0587
1200	0.0405	24.691	1457.5	1640.4	2.0917
1300	0.03819	26.185	1500.4	1694.4	2.1233
1400	0.03613	27.678	1544.3	1749.3	2.1536
1500	0.034282	29.170	1589.1	1805.2	2.1829
1600	0.032614	30.662	1634.8	1861.9	2.2111

Table DE.3G $P = 60$ psia ($T_{sat} = 292.68°F$)

T, °F	ρ, lbm/ft³	v, ft³/lbm	u, Btu/lbm	h, Btu/lbm	s, Btu/lbm·R
Sat.	0.13935	7.1761	1098.9	1178.6	1.6454
320	0.13358	7.4863	1110.3	1193.5	1.6648
360	0.12617	7.9259	1126.2	1214.3	1.6908
400	0.11969	8.3548	1141.7	1234.5	1.7149
440	0.11394	8.7766	1156.9	1254.4	1.7376
500	0.10638	9.4004	1179.6	1284.0	1.7694
600	0.095918	10.426	1217.3	1333.1	1.8181
700	0.087412	11.440	1255.4	1382.5	1.8626
800	0.080332	12.448	1294.1	1432.4	1.9039
900	0.074335	13.453	1333.6	1483.1	1.9426
1000	0.069184	14.454	1373.9	1534.5	1.9791
1100	0.064708	15.454	1415.1	1586.8	2.0137
1200	0.060781	16.452	1457.2	1640.0	2.0468
1300	0.057307	17.450	1500.1	1694.0	2.0784
1400	0.054211	18.446	1544.1	1749.0	2.1088
1500	0.051434	19.442	1588.9	1804.9	2.1380
1600	0.048929	20.438	1634.6	1861.7	2.1663
1700	0.046657	21.433	1681.3	1919.4	2.1937
1800	0.044588	22.428	1728.8	1978.0	2.2202

Table DE.3H $P = 80$ psia ($T_{sat} = 312.02°F$)

T, °F	ρ, lbm/ft³	v, ft³/lbm	u, Btu/lbm	h, Btu/lbm	s, Btu/lbm·R
Sat.	0.18271	5.4730	1103.1	1184.1	1.6223
320	0.18037	5.5440	1106.6	1188.7	1.6282
360	0.16985	5.8875	1123.4	1210.7	1.6556
400	0.16081	6.2186	1139.5	1231.6	1.6806
440	0.15286	6.5420	1155.1	1252.0	1.7038
500	0.14250	7.0176	1178.1	1282.1	1.7362
600	0.12829	7.7951	1216.2	1331.7	1.7854
700	0.11680	8.5616	1254.6	1381.4	1.8302
800	0.10728	9.3217	1293.5	1431.6	1.8717
900	0.099227	10.078	1333.1	1482.4	1.9105
1000	0.092325	10.831	1373.5	1534.0	1.9471
1100	0.086334	11.583	1414.7	1586.3	1.9818
1200	0.081083	12.333	1456.9	1639.6	2.0148
1300	0.076440	13.082	1499.9	1693.7	2.0465
1400	0.072304	13.831	1543.8	1748.7	2.0769
1500	0.068595	14.578	1588.7	1804.6	2.1062
1600	0.065250	15.326	1634.4	1861.5	2.1345
1700	0.062218	16.073	1681.1	1919.2	2.1619
1800	0.059456	16.819	1728.6	1977.8	2.1884

Table DE.3I $P = 100$ psia ($T_{sat} = 327.81°F$)

T, °F	ρ, lbm/ft³	v, ft³/lbm	u, Btu/lbm	h, Btu/lbm	s, Btu/lbm·R
Sat.	0.22560	4.4326	1106.2	1188.3	1.6043
360	0.21447	4.6627	1120.5	1206.9	1.6274
400	0.20260	4.9359	1137.2	1228.6	1.6532
440	0.19228	5.2006	1153.2	1249.5	1.6770
500	0.17897	5.5876	1176.7	1280.1	1.7100
600	0.16086	6.2166	1215.2	1330.3	1.7598
700	0.14632	6.8344	1253.8	1380.4	1.8049
800	0.13431	7.4457	1292.9	1430.8	1.8466
900	0.12418	8.0530	1332.6	1481.7	1.8855
1000	0.11551	8.6575	1373.1	1533.4	1.9222
1100	0.10799	9.2602	1414.4	1585.9	1.9569
1200	0.10141	9.8614	1456.5	1639.2	1.9900
1300	0.095587	10.462	1499.6	1693.3	2.0217
1400	0.090407	11.061	1543.6	1748.4	2.0522
1500	0.085763	11.660	1588.5	1804.4	2.0815
1600	0.081577	12.258	1634.2	1861.2	2.1098
1700	0.077783	12.856	1680.9	1919.0	2.1372
1800	0.074327	13.454	1728.5	1977.6	2.1637

Table DE.3J $P = 120$ psia ($T_{sat} = 341.25°F$)

T, °F	ρ, lbm/ft³	v, ft³/lbm	u, Btu/lbm	h, Btu/lbm	s, Btu/lbm·R
Sat.	0.26818	3.7288	1108.7	1191.5	1.5894
350	0.26431	3.7834	1112.8	1196.9	1.5961
400	0.24511	4.0799	1134.8	1225.4	1.6303
450	0.22929	4.3612	1155.3	1252.2	1.6606
500	0.21580	4.6340	1175.2	1278.2	1.6883
550	0.20404	4.9010	1194.7	1303.6	1.7142
600	0.19364	5.1642	1214.2	1328.9	1.7387
700	0.17597	5.6829	1253.0	1379.3	1.7841
800	0.16142	6.1950	1292.3	1429.9	1.8260
900	0.14919	6.7030	1332.1	1481.1	1.8650
1000	0.13873	7.2083	1372.7	1532.8	1.9018
1100	0.12967	7.7117	1414.0	1585.4	1.9366
1200	0.12175	8.2137	1456.2	1638.8	1.9697
1300	0.11475	8.7146	1499.3	1693.0	2.0015
1400	0.10852	9.2148	1543.3	1748.1	2.0319
1500	0.10294	9.7144	1588.2	1804.1	2.0613
1600	0.09791	10.213	1634.0	1861.0	2.0896
1700	0.093351	10.712	1680.7	1918.8	2.1170
1800	0.089201	11.211	1728.3	1977.4	2.1435

Table DE.3K $P = 140$ psia ($T_{sat} = 353.03°F$)

T, °F	ρ, lbm/ft³	v, ft³/lbm	u, Btu/lbm	h, Btu/lbm	s, Btu/lbm·R
Sat.	0.31055	3.2201	1110.7	1194.2	1.5768
360	0.30690	3.2584	1114.1	1198.6	1.5822
400	0.28839	3.4676	1132.3	1222.2	1.6103
450	0.26920	3.7147	1153.4	1249.7	1.6415
500	0.25300	3.9525	1173.7	1276.1	1.6697
550	0.23898	4.1845	1193.5	1302.0	1.6960
600	0.22663	4.4124	1213.1	1327.5	1.7207
700	0.20575	4.8603	1252.2	1378.2	1.7664
800	0.18862	5.3016	1291.6	1429.1	1.8085
900	0.17425	5.7388	1331.6	1480.4	1.8476
1000	0.16199	6.1732	1372.2	1532.3	1.8845
1100	0.15139	6.6056	1413.7	1584.9	1.9194
1200	0.14211	7.0367	1455.9	1638.4	1.9526
1300	0.13393	7.4668	1499.1	1692.6	1.9843
1400	0.12665	7.8960	1543.1	1747.8	2.0148
1500	0.12012	8.3247	1588.0	1803.8	2.0442
1600	0.11425	8.7529	1633.9	1860.8	2.0725
1700	0.10892	9.1807	1680.6	1918.6	2.0999
1800	0.10408	9.6082	1728.2	1977.2	2.1264

Table DE.3L $P = 160$ psia ($T_{sat} = 363.54°F$)

T, °F	ρ, lbm/ft³	v, ft³/lbm	u, Btu/lbm	h, Btu/lbm	s, Btu/lbm·R
Sat.	0.35278	2.8347	1112.3	1196.3	1.5657
400	0.33250	3.0076	1129.7	1218.8	1.5925
450	0.30966	3.2293	1151.5	1247.2	1.6245
500	0.29060	3.4412	1172.1	1274.1	1.6534
550	0.27421	3.6469	1192.2	1300.3	1.6800
600	0.25985	3.8484	1212.1	1326.1	1.7049
700	0.23566	4.2434	1251.5	1377.2	1.7510
800	0.21591	4.6316	1291.0	1428.2	1.7932
900	0.19938	5.0156	1331.1	1479.7	1.8325
1000	0.18530	5.3968	1371.8	1531.7	1.8695
1100	0.17313	5.7761	1413.3	1584.4	1.9044
1200	0.16250	6.1540	1455.6	1638.0	1.9376
1300	0.15312	6.5309	1498.8	1692.3	1.9694
1400	0.14478	6.9069	1542.9	1747.5	1.9999
1500	0.13732	7.2824	1587.8	1803.6	2.0293
1600	0.13059	7.6574	1633.7	1860.5	2.0577
1700	0.12450	8.0320	1680.4	1918.4	2.0851
1800	0.11896	8.4063	1728.0	1977.1	2.1116

Table DE.3M $P = 180$ psia ($T_{sat} = 373.07°F$)

T, °F	ρ, lbm/ft³	v, ft³/lbm	u, Btu/lbm	h, Btu/lbm	s, Btu/lbm·R
Sat.	0.39492	2.5321	1113.7	1198.1	1.5559
400	0.37750	2.6490	1127.0	1215.3	1.5763
450	0.35070	2.8514	1149.5	1244.5	1.6093
500	0.32859	3.0433	1170.6	1272.0	1.6387
550	0.30973	3.2286	1191.0	1298.6	1.6657
600	0.29328	3.4097	1211.0	1324.6	1.6909
700	0.26571	3.7635	1250.7	1376.1	1.7373
800	0.24329	4.1104	1290.4	1427.4	1.7798
900	0.22456	4.4531	1330.6	1479.0	1.8192
1000	0.20864	4.7929	1371.4	1531.1	1.8562
1100	0.19490	5.1309	1412.9	1584.0	1.8912
1200	0.18290	5.4674	1455.3	1637.5	1.9245
1300	0.17233	5.8029	1498.5	1691.9	1.9563
1400	0.16293	6.1377	1542.6	1747.2	1.9868
1500	0.15452	6.4718	1587.6	1803.3	2.0162
1600	0.14694	6.8054	1633.5	1860.3	2.0446
1700	0.14008	7.1386	1680.2	1918.2	2.0720
1800	0.13384	7.4716	1727.8	1976.9	2.0986

Table DE.3N $P = 200$ psia ($T_{sat} = 381.80°F$)

T, °F	ρ, lbm/ft³	v, ft³/lbm	u, Btu/lbm	h, Btu/lbm	s, Btu/lbm·R
Sat.	0.43703	2.2882	1114.8	1199.6	1.5470
400	0.42347	2.3615	1124.2	1211.7	1.5613
450	0.39234	2.5488	1147.4	1241.8	1.5954
500	0.36701	2.7247	1169.0	1269.9	1.6254
550	0.34555	2.8939	1189.7	1296.8	1.6528
600	0.32695	3.0586	1209.9	1323.2	1.6782
700	0.29590	3.3795	1249.8	1375.0	1.7250
800	0.27075	3.6934	1289.8	1426.6	1.7676
900	0.24981	4.0030	1330.1	1478.3	1.8072
1000	0.23203	4.3098	1371.0	1530.6	1.8443
1100	0.21670	4.6147	1412.6	1583.5	1.8793
1200	0.20333	4.9182	1455.0	1637.1	1.9127
1300	0.19155	5.2206	1498.3	1691.6	1.9445
1400	0.18109	5.5222	1542.4	1746.9	1.9751
1500	0.17173	5.8232	1587.4	1803.1	2.0045
1600	0.16330	6.1238	1633.3	1860.1	2.0329
1700	0.15567	6.4239	1680.1	1918.0	2.0603
1800	0.14873	6.7237	1727.7	1976.7	2.0869

Table DE.3O $P = 225$ psia ($T_{sat} = 391.80°F$)

T, °F	ρ, lbm/ft³	v, ft³/lbm	u, Btu/lbm	h, Btu/lbm	s, Btu/lbm·R
Sat.	0.48964	2.0423	1116.0	1201.1	1.5370
400	0.48244	2.0728	1120.5	1206.8	1.5437
450	0.44530	2.2457	1144.8	1238.4	1.5794
500	0.41564	2.4059	1166.9	1267.2	1.6102
550	0.39077	2.5590	1188.0	1294.6	1.6381
600	0.36935	2.7075	1208.5	1321.3	1.6639
700	0.33383	2.9955	1248.8	1373.6	1.7111
800	0.30521	3.2765	1289.0	1425.5	1.7540
900	0.28145	3.5530	1329.4	1477.5	1.7937
1000	0.26132	3.8268	1370.4	1529.9	1.8309
1100	0.24399	4.0985	1412.1	1582.9	1.8660
1200	0.22889	4.3689	1454.6	1636.6	1.8994
1300	0.21560	4.6382	1497.9	1691.2	1.9313
1400	0.20380	4.9068	1542.1	1746.5	1.9619
1500	0.19325	5.1747	1587.1	1802.7	1.9914
1600	0.18375	5.4421	1633.1	1859.8	2.0198
1700	0.17516	5.7092	1679.8	1917.7	2.0472
1800	0.16734	5.9759	1727.5	1976.5	2.0738

Table DE.3P $P = 250$ psia ($T_{sat} = 400.97°F$)

T, °F	ρ, lbm/ft³	v, ft³/lbm	u, Btu/lbm	h, Btu/lbm	s, Btu/lbm·R
Sat.	0.54230	1.8440	1117.0	1202.4	1.5280
450	0.49933	2.0027	1142.1	1234.8	1.5647
500	0.46498	2.1506	1164.9	1264.4	1.5964
550	0.43650	2.2910	1186.3	1292.4	1.6248
600	0.41213	2.4264	1207.1	1319.5	1.6510
650	0.39085	2.5586	1227.6	1346.0	1.6755
700	0.37198	2.6883	1247.8	1372.3	1.6986
800	0.33981	2.9428	1288.2	1424.4	1.7417
900	0.31319	3.1930	1328.8	1476.6	1.7816
1000	0.29067	3.4403	1369.9	1529.2	1.8189
1100	0.27133	3.6856	1411.7	1582.3	1.8541
1200	0.25449	3.9295	1454.2	1636.1	1.8876
1300	0.23967	4.1724	1497.6	1690.7	1.9195
1400	0.22653	4.4144	1541.8	1746.2	1.9501
1500	0.21478	4.6559	1586.9	1802.4	1.9796
1600	0.20421	4.8968	1632.8	1859.5	2.0080
1700	0.19465	5.1374	1679.6	1917.5	2.0355
1800	0.18595	5.3777	1727.3	1976.3	2.0621

Table DE.3Q $P = 275$ psia ($T_{sat} = 409.45°F$)

T, °F	ρ, lbm/ft³	v, ft³/lbm	u, Btu/lbm	h, Btu/lbm	s, Btu/lbm·R
Sat.	0.59503	1.6806	1117.8	1203.4	1.5197
450	0.55450	1.8034	1139.3	1231.1	1.5510
500	0.51507	1.9415	1162.7	1261.6	1.5836
550	0.48275	2.0715	1184.6	1290.1	1.6126
600	0.45529	2.1964	1205.7	1317.6	1.6391
650	0.43143	2.3179	1226.4	1344.4	1.6639
700	0.41036	2.4369	1246.8	1370.9	1.6872
800	0.37455	2.6699	1287.4	1423.3	1.7306
900	0.34502	2.8984	1328.1	1475.7	1.7706
1000	0.32010	3.1241	1369.4	1528.5	1.8080
1100	0.29871	3.3477	1411.2	1581.7	1.8433
1200	0.28011	3.5700	1453.8	1635.6	1.8768
1300	0.26377	3.7912	1497.2	1690.3	1.9088
1400	0.24928	4.0116	1541.5	1745.8	1.9395
1500	0.23633	4.2314	1586.6	1802.1	1.9690
1600	0.22468	4.4507	1632.6	1859.2	1.9974
1700	0.21415	4.6696	1679.4	1917.2	2.0249
1800	0.20457	4.8882	1727.1	1976.0	2.0515

Table DE.3R $P = 300$ psia ($T_{sat} = 417.35°F$)

T, °F	ρ, lbm/ft³	v, ft³/lbm	u, Btu/lbm	h, Btu/lbm	s, Btu/lbm·R
Sat.	0.64787	1.5435	1118.4	1204.1	1.5121
450	0.61092	1.6369	1136.3	1227.3	1.5380
500	0.56594	1.7670	1160.5	1258.7	1.5716
550	0.52954	1.8884	1182.9	1287.8	1.6012
600	0.49885	2.0046	1204.3	1315.7	1.6282
650	0.47232	2.1172	1225.2	1342.8	1.6532
700	0.44897	2.2273	1245.8	1369.5	1.6767
800	0.40944	2.4424	1286.6	1422.3	1.7204
900	0.37694	2.6529	1327.5	1474.9	1.7605
1000	0.34958	2.8605	1368.8	1527.7	1.7981
1100	0.32614	3.0662	1410.8	1581.1	1.8334
1200	0.30577	3.2704	1453.5	1635.1	1.8670
1300	0.28789	3.4735	1496.9	1689.9	1.8990
1400	0.27204	3.6759	1541.2	1745.4	1.9297
1500	0.25789	3.8776	1586.3	1801.8	1.9592
1600	0.24516	4.0789	1632.3	1858.9	1.9877
1700	0.23366	4.2798	1679.2	1917.0	2.0152
1800	0.22320	4.4803	1726.9	1975.8	2.0418

Table DE.3S $P = 350$ psia ($T_{sat} = 431.74°F$)

T, °F	ρ, lbm/ft³	v, ft³/lbm	u, Btu/lbm	h, Btu/lbm	s, Btu/lbm·R
Sat.	0.75400	1.3263	1119.3	1205.2	1.4984
450	0.72788	1.3739	1130.1	1219.1	1.5138
500	0.67021	1.4921	1156.0	1252.7	1.5497
550	0.62483	1.6004	1179.4	1283.1	1.5806
600	0.58720	1.7030	1201.4	1311.8	1.6084
650	0.55501	1.8018	1222.7	1339.5	1.6339
700	0.52690	1.8979	1243.7	1366.7	1.6579
800	0.47966	2.0848	1285.0	1420.1	1.7021
900	0.44109	2.2671	1326.2	1473.1	1.7426
1000	0.40876	2.4464	1367.8	1526.3	1.7803
1100	0.38114	2.6237	1409.9	1579.9	1.8158
1200	0.35720	2.7996	1452.7	1634.1	1.8495
1300	0.33620	2.9744	1496.2	1689.0	1.8816
1400	0.31762	3.1484	1540.6	1744.6	1.9124
1500	0.30105	3.3218	1585.8	1801.1	1.9420
1600	0.28615	3.4947	1631.9	1858.4	1.9704
1700	0.27269	3.6672	1678.8	1916.5	1.9980
1800	0.26046	3.8393	1726.5	1975.4	2.0247

Table DE.3T $P = 400$ psia ($T_{sat} = 444.62°F$)

T, °F	ρ, lbm/ft³	v, ft³/lbm	u, Btu/lbm	h, Btu/lbm	s, Btu/lbm·R
Sat.	0.86087	1.1616	1119.8	1205.8	1.4862
450	0.85127	1.1747	1123.2	1210.2	1.4911
500	0.77816	1.2851	1151.2	1246.4	1.5298
550	0.72256	1.3840	1175.7	1278.2	1.5621
600	0.67726	1.4765	1198.4	1307.8	1.5907
650	0.63897	1.5650	1220.2	1336.2	1.6169
700	0.60579	1.6507	1241.5	1363.8	1.6413
800	0.55048	1.8166	1283.3	1417.9	1.6860
900	0.50563	1.9777	1324.9	1471.4	1.7269
1000	0.46821	2.1358	1366.7	1524.9	1.7649
1100	0.43633	2.2919	1409.0	1578.7	1.8005
1200	0.40875	2.4465	1451.9	1633.1	1.8343
1300	0.38461	2.6000	1495.6	1688.1	1.8665
1400	0.36327	2.7527	1540.0	1743.9	1.8973
1500	0.34425	2.9048	1585.3	1800.4	1.9270
1600	0.32717	3.0565	1631.4	1857.8	1.9555
1700	0.31175	3.2077	1678.4	1915.9	1.9831
1800	0.29774	3.3586	1726.1	1974.9	2.0098

Table DE.3U $P = 450$ psia ($T_{sat} = 456.31°F$)

T, °F	ρ, lbm/ft³	v, ft³/lbm	u, Btu/lbm	h, Btu/lbm	s, Btu/lbm·R
Sat.	0.96864	1.0324	1119.9	1206.0	1.4752
500	0.89025	1.1233	1146.2	1239.8	1.5113
550	0.82292	1.2152	1171.8	1273.1	1.5452
600	0.76916	1.3001	1195.4	1303.7	1.5748
650	0.72426	1.3807	1217.7	1332.8	1.6016
700	0.68569	1.4584	1239.4	1360.9	1.6264
800	0.62191	1.6079	1281.7	1415.7	1.6717
900	0.57057	1.7526	1323.6	1469.6	1.7129
1000	0.52792	1.8942	1365.6	1523.4	1.7511
1100	0.49170	2.0338	1408.1	1577.5	1.7870
1200	0.46044	2.1718	1451.1	1632.1	1.8209
1300	0.43312	2.3088	1494.9	1687.3	1.8531
1400	0.40900	2.4450	1539.4	1743.1	1.8840
1500	0.38751	2.5806	1584.7	1799.8	1.9137
1600	0.36823	2.7157	1630.9	1857.2	1.9423
1700	0.35083	2.8503	1677.9	1915.4	1.9699
1800	0.33504	2.9847	1725.8	1974.5	1.9966

Table DE.3V $P = 500$ psia ($T_{sat} = 467.04°F$)

T, °F	ρ, lbm/ft³	v, ft³/lbm	u, Btu/lbm	h, Btu/lbm	s, Btu/lbm·R
Sat.	1.0774	0.92815	1119.8	1205.8	1.4652
500	1.0070	0.99303	1140.8	1232.8	1.4938
550	0.92615	1.0797	1167.9	1267.8	1.5295
600	0.86300	1.1588	1192.2	1299.5	1.5601
650	0.81095	1.2331	1215.1	1329.3	1.5876
700	0.76664	1.3044	1237.2	1357.9	1.6128
800	0.69398	1.4410	1280.0	1413.4	1.6588
900	0.63592	1.5725	1322.3	1467.9	1.7003
1000	0.58791	1.7009	1364.5	1522.0	1.7387
1100	0.54727	1.8273	1407.1	1576.3	1.7747
1200	0.51227	1.9521	1450.3	1631.1	1.8088
1300	0.48172	2.0759	1494.2	1686.4	1.8411
1400	0.45479	2.1988	1538.8	1742.4	1.8721
1500	0.43082	2.3212	1584.2	1799.1	1.9018
1600	0.40933	2.4430	1630.4	1856.6	1.9304
1700	0.38995	2.5645	1677.5	1914.9	1.9581
1800	0.37236	2.6856	1725.4	1974.0	1.9848

Table DE.3W $P = 600$ psia ($T_{sat} = 486.24°F$)

T, °F	ρ, lbm/ft³	v, ft³/lbm	u, Btu/lbm	h, Btu/lbm	s, Btu/lbm·R
Sat.	1.2984	0.77015	1119.1	1204.6	1.4473
500	1.2575	0.79526	1129.0	1217.3	1.4606
550	1.1423	0.87542	1159.4	1256.7	1.5006
600	1.0570	0.94605	1185.7	1290.8	1.5336
650	0.98880	1.0113	1209.8	1322.1	1.5625
700	0.93183	1.0732	1232.7	1351.9	1.5888
800	0.84007	1.1904	1276.7	1408.9	1.6359
900	0.76788	1.3023	1319.6	1464.3	1.6782
1000	0.70874	1.4110	1362.3	1519.1	1.7171
1100	0.65898	1.5175	1405.3	1573.9	1.7534
1200	0.61633	1.6225	1448.8	1629.0	1.7877
1300	0.57922	1.7264	1492.8	1684.6	1.8202
1400	0.54658	1.8296	1537.6	1740.9	1.8513
1500	0.51759	1.9320	1583.2	1797.8	1.8811
1600	0.49163	2.0340	1629.5	1855.5	1.9098
1700	0.46825	2.1356	1676.6	1913.9	1.9375
1800	0.44705	2.2369	1724.6	1973.1	1.9643

Table DE.3X $P = 700$ psia ($T_{sat} = 503.13°F$)

T, °F	ρ, lbm/ft³	v, ft³/lbm	u, Btu/lbm	h, Btu/lbm	s, Btu/lbm·R
Sat.	1.5247	0.65586	1117.7	1202.7	1.4315
550	1.3736	0.72799	1150.2	1244.6	1.4740
600	1.2605	0.79332	1178.7	1281.5	1.5097
650	1.1731	0.85242	1204.2	1314.7	1.5403
700	1.1017	0.90769	1228.1	1345.7	1.5677
800	0.98888	1.0112	1273.2	1404.3	1.6162
900	0.90154	1.1092	1316.9	1460.7	1.6592
1000	0.83070	1.2038	1360.1	1516.2	1.6986
1100	0.77148	1.2962	1403.5	1571.5	1.7353
1200	0.72094	1.3871	1447.2	1627.0	1.7697
1300	0.67712	1.4769	1491.5	1682.9	1.8025
1400	0.63865	1.5658	1536.4	1739.4	1.8337
1500	0.60456	1.6541	1582.1	1796.5	1.8636
1600	0.57408	1.7419	1628.5	1854.3	1.8924
1700	0.54665	1.8293	1675.8	1912.9	1.9201
1800	0.52181	1.9164	1723.8	1972.2	1.9470

Table DE.3Y $P = 800$ psia ($T_{sat} = 518.27°F$)

T, °F	ρ, lbm/ft^3	v, ft^3/lbm	u, Btu/lbm	h, Btu/lbm	s, Btu/lbm·R
Sat.	1.7569	0.56917	1115.8	1200.1	1.4171
550	1.6237	0.61586	1140.1	1231.3	1.4486
600	1.4750	0.67799	1171.3	1271.7	1.4877
650	1.3647	0.73279	1198.4	1306.9	1.5201
700	1.2766	0.78330	1223.3	1339.3	1.5487
750	1.2033	0.83102	1246.9	1370.0	1.5746
800	1.1405	0.87678	1269.7	1399.6	1.5986
900	1.0370	0.96433	1314.2	1457.0	1.6425
1000	0.95383	1.0484	1357.9	1513.2	1.6823
1100	0.88477	1.1302	1401.6	1569.0	1.7193
1200	0.82610	1.2105	1445.6	1624.9	1.7541
1300	0.77540	1.2897	1490.1	1681.1	1.7869
1400	0.73101	1.3680	1535.2	1737.8	1.8183
1500	0.69173	1.4456	1581.0	1795.2	1.8483
1600	0.65668	1.5228	1627.6	1853.2	1.8772
1700	0.62516	1.5996	1674.9	1911.9	1.9050
1800	0.59664	1.6761	1723.0	1971.3	1.9319

Table DE.3Z $P = 1000$ psia ($T_{sat} = 544.65°F$)

T, °F	ρ, lbm/ft^3	v, ft^3/lbm	u, Btu/lbm	h, Btu/lbm	s, Btu/lbm·R
Sat.	2.2419	0.44605	1110.8	1193.4	1.3916
550	2.2039	0.45375	1116.0	1200.0	1.3981
600	1.9444	0.51431	1154.9	1250.1	1.4466
650	1.7727	0.56410	1185.9	1290.3	1.4837
700	1.6435	0.60844	1213.2	1325.8	1.5151
750	1.5398	0.64944	1238.4	1358.7	1.5428
800	1.4531	0.68821	1262.5	1389.9	1.5681
900	1.3134	0.76136	1308.6	1449.6	1.6137
1000	1.2037	0.83077	1353.4	1507.2	1.6546
1100	1.1138	0.89782	1397.8	1564.1	1.6923
1200	1.0381	0.96327	1442.4	1620.8	1.7275
1300	0.97318	1.0276	1487.3	1677.6	1.7608
1400	0.91658	1.0910	1532.8	1734.8	1.7924
1500	0.86670	1.1538	1578.9	1792.5	1.8226
1600	0.82230	1.2161	1625.7	1850.9	1.8516
1700	0.78248	1.2780	1673.2	1909.9	1.8796
1800	0.74652	1.3396	1721.5	1969.5	1.9066

Table DE.3AA $P = 1250$ psia ($T_{sat} = 572.45°F$)

T, °F	ρ, lbm/ft^3	v, ft^3/lbm	u, Btu/lbm	h, Btu/lbm	s, Btu/lbm·R
Sat.	2.8942	0.34551	1102.7	1182.7	1.3633
600	2.6389	0.37894	1130.3	1218.0	1.3970
650	2.3418	0.42703	1168.3	1267.2	1.4424
700	2.1397	0.46734	1199.5	1307.7	1.4781
750	1.9863	0.50344	1227.2	1343.8	1.5086
800	1.8627	0.53687	1253.0	1377.3	1.5358
900	1.6701	0.59876	1301.4	1440.0	1.5837
1000	1.5231	0.65655	1347.6	1499.6	1.6260
1100	1.4048	0.71184	1393.1	1557.8	1.6646
1200	1.3064	0.76545	1438.4	1615.5	1.7005
1300	1.2227	0.81788	1483.8	1673.2	1.7342
1400	1.1502	0.86944	1529.7	1731.0	1.7661
1500	1.0866	0.92033	1576.2	1789.2	1.7966
1600	1.0302	0.97072	1623.3	1848.0	1.8259
1700	0.97972	1.0207	1671.1	1907.3	1.8540
1800	0.93427	1.0704	1719.5	1967.3	1.8811

Table DE.3BB $P = 1500$ psia ($T_{sat} = 596.26°F$)

T, °F	ρ, lbm/ft^3	v, ft^3/lbm	u, Btu/lbm	h, Btu/lbm	s, Btu/lbm·R
Sat.	3.6105	0.27697	1092.9	1169.8	1.3372
600	3.5475	0.28189	1097.9	1176.2	1.3432
650	3.0021	0.33310	1148.0	1240.5	1.4026
700	2.6883	0.37198	1184.4	1287.7	1.4443
750	2.4670	0.40534	1215.2	1327.8	1.4781
800	2.2962	0.43550	1243.0	1364.0	1.5075
850	2.1572	0.46356	1269.1	1397.8	1.5338
900	2.0402	0.49014	1294.0	1430.1	1.5580
1000	1.8508	0.54031	1341.8	1491.8	1.6018
1100	1.7012	0.58780	1388.3	1551.5	1.6414
1200	1.5784	0.63355	1434.3	1610.3	1.6779
1300	1.4748	0.67808	1480.3	1668.7	1.7121
1400	1.3856	0.72172	1526.7	1727.1	1.7444
1500	1.3077	0.76469	1573.5	1785.9	1.7751
1600	1.2389	0.80714	1620.9	1845.1	1.8046
1700	1.1776	0.84919	1668.9	1904.8	1.8329
1800	1.1225	0.89090	1717.6	1965.0	1.8602

Table DE.3CC $P = 1750$ psia ($T_{sat} = 617.17°F$)

T, °F	ρ, lbm/ft³	v, ft³/lbm	u, Btu/lbm	h, Btu/lbm	s, Btu/lbm·R
Sat.	4.4088	0.22682	1081.2	1154.7	1.3121
650	3.8034	0.26292	1123.5	1208.7	1.3616
700	3.3056	0.30252	1167.5	1265.6	1.4118
750	2.9891	0.33455	1202.2	1310.7	1.4499
800	2.7574	0.36266	1232.5	1350.0	1.4818
850	2.5750	0.38835	1260.2	1386.0	1.5098
900	2.4249	0.41238	1286.3	1419.9	1.5352
1000	2.1873	0.45719	1335.8	1483.9	1.5806
1100	2.0033	0.49917	1383.4	1545.1	1.6212
1200	1.8542	0.53932	1430.2	1604.9	1.6584
1300	1.7294	0.57822	1476.8	1664.2	1.6930
1400	1.6228	0.61621	1523.6	1723.3	1.7257
1500	1.5302	0.65352	1570.8	1782.6	1.7568
1600	1.4486	0.69031	1618.5	1842.2	1.7864
1700	1.3761	0.72668	1666.7	1902.2	1.8149
1800	1.3111	0.76273	1715.6	1962.8	1.8423

Table DE.3DD $P = 2000$ psia ($T_{sat} = 635.85°F$)

T, °F	ρ, lbm/ft³	v, ft³/lbm	u, Btu/lbm	h, Btu/lbm	s, Btu/lbm·R
Sat.	5.3156	0.18813	1067.5	1137.2	1.2872
650	4.8576	0.20586	1092.1	1168.4	1.3155
700	4.0171	0.24894	1148.4	1240.6	1.3793
750	3.5621	0.28074	1188.2	1292.1	1.4228
800	3.2507	0.30763	1221.3	1335.2	1.4577
850	3.0149	0.33169	1250.9	1373.7	1.4877
900	2.8256	0.35390	1278.3	1409.4	1.5145
1000	2.5330	0.39479	1329.6	1475.8	1.5616
1100	2.3113	0.43266	1378.4	1538.6	1.6032
1200	2.1339	0.46864	1426.0	1599.6	1.6411
1300	1.9868	0.50332	1473.3	1659.7	1.6763
1400	1.8619	0.53708	1520.5	1719.4	1.7093
1500	1.7539	0.57015	1568.1	1779.2	1.7406
1600	1.6592	0.60269	1616.0	1839.3	1.7705
1700	1.5753	0.63481	1664.6	1899.7	1.7992
1800	1.5002	0.66660	1713.7	1960.6	1.8267

Table DE.3EE $P = 2500$ psia ($T_{sat} = 668.17°F$)

T, °F	ρ, lbm/ft³	v, ft³/lbm	u, Btu/lbm	h, Btu/lbm	s, Btu/lbm·R
Sat.	7.6508	0.13071	1031.8	1092.3	1.2338
700	5.9350	0.16849	1099.1	1177.1	1.3081
750	4.9195	0.20327	1155.7	1249.8	1.3695
800	4.3575	0.22949	1196.7	1302.9	1.4125
850	3.9724	0.25174	1231.0	1347.5	1.4473
900	3.6812	0.27165	1261.6	1387.3	1.4772
950	3.4482	0.29001	1290.0	1424.2	1.5038
1000	3.2546	0.30726	1316.9	1459.2	1.5282
1100	2.9456	0.33949	1368.2	1525.4	1.5720
1200	2.7052	0.36966	1417.6	1588.7	1.6114
1300	2.5096	0.39847	1466.1	1650.5	1.6476
1400	2.3457	0.42631	1514.3	1711.7	1.6814
1500	2.2053	0.45344	1562.6	1772.5	1.7132
1600	2.0832	0.48004	1611.2	1833.4	1.7436
1700	1.9755	0.50621	1660.2	1894.6	1.7725
1800	1.8795	0.53204	1709.8	1956.1	1.8004

Table DE.3FF $P = 3000$ psia ($T_{sat} = 695.41°F$)

T, °F	ρ, lbm/ft³	v, ft³/lbm	u, Btu/lbm	h, Btu/lbm	s, Btu/lbm·R
Sat.	11.812	0.084657	970.58	1017.6	1.1597
700	10.165	0.098374	1006.0	1060.6	1.1969
750	6.7386	0.14840	1114.8	1197.2	1.3127
800	5.6815	0.17601	1168.3	1266.1	1.3685
850	5.0578	0.19771	1209.0	1318.8	1.4096
900	4.6212	0.21639	1243.6	1363.8	1.4433
950	4.2880	0.23321	1274.7	1404.3	1.4726
1000	4.0200	0.24876	1303.7	1441.9	1.4988
1100	3.6060	0.27732	1357.8	1511.8	1.5451
1200	3.2931	0.30367	1409.0	1577.7	1.5861
1300	3.0434	0.32858	1458.8	1641.3	1.6233
1400	2.8370	0.35249	1508.0	1703.8	1.6579
1500	2.6619	0.37567	1557.1	1765.8	1.6903
1600	2.5107	0.39830	1606.3	1827.6	1.7211
1700	2.3781	0.42050	1655.9	1889.5	1.7504
1800	2.2606	0.44236	1705.8	1951.6	1.7785

Table DE.3GG $P = 3500$ psia

T, °F	ρ, lbm/ft³	v, ft³/lbm	u, Btu/lbm	h, Btu/lbm	s, Btu/lbm·R
650	40.123	0.024924	664.18	680.33	0.86378
700	32.624	0.030653	760.51	780.37	0.95177
750	9.5603	0.10460	1058.3	1126.1	1.2443
800	7.3320	0.13639	1135.0	1223.4	1.3233
850	6.3102	0.15847	1184.6	1287.3	1.3730
900	5.6630	0.17659	1224.2	1338.7	1.4115
950	5.1961	0.19245	1258.6	1383.3	1.4438
1000	4.8341	0.20687	1289.9	1423.9	1.4721
1100	4.2939	0.23289	1347.0	1497.9	1.5212
1200	3.8981	0.25654	1400.2	1566.5	1.5638
1300	3.5884	0.27868	1451.4	1632.0	1.6021
1400	3.3358	0.29978	1501.7	1696.0	1.6375
1500	3.1236	0.32014	1551.6	1759.1	1.6705
1600	2.9417	0.33994	1601.4	1821.7	1.7017
1700	2.7831	0.35931	1651.5	1884.4	1.7314
1800	2.6432	0.37833	1701.9	1947.1	1.7598

Table DE.3HH $P = 4000$ psia

T, °F	ρ, lbm/ft³	v, ft³/lbm	u, Btu/lbm	h, Btu/lbm	s, Btu/lbm·R
650	40.843	0.024484	658.37	676.51	0.85827
700	34.828	0.028713	742.81	764.08	0.93536
750	15.698	0.063703	962.70	1009.9	1.1417
800	9.5056	0.10520	1095.0	1172.9	1.2743
850	7.7834	0.12848	1157.4	1252.6	1.3364
900	6.8276	0.14647	1203.3	1311.8	1.3808
950	6.1822	0.16175	1241.6	1361.4	1.4166
1000	5.7018	0.17538	1275.4	1405.3	1.4473
1100	5.0109	0.19957	1336.0	1483.8	1.4993
1200	4.5206	0.22121	1391.3	1555.1	1.5437
1300	4.1446	0.24128	1443.9	1622.7	1.5832
1400	3.8420	0.26028	1495.3	1688.1	1.6193
1500	3.5903	0.27853	1546.0	1752.3	1.6530
1600	3.3761	0.29620	1596.5	1815.9	1.6846
1700	3.1904	0.31344	1647.1	1879.3	1.7147
1800	3.0272	0.33033	1698.0	1942.6	1.7434

Table DE.3II $P = 5000$ psia

T, °F	ρ, lbm/ft³	v, ft³/lbm	u, Btu/lbm	h, Btu/lbm	s, Btu/lbm·R
650	42.037	0.023788	648.71	670.73	0.84904
700	37.345	0.026777	722.26	747.05	0.91627
750	29.647	0.033730	822.38	853.61	1.0061
800	16.845	0.059365	987.53	1042.5	1.1589
850	11.695	0.085508	1093.1	1172.3	1.2602
900	9.6243	0.10390	1156.7	1252.9	1.3207
950	8.4295	0.11863	1204.7	1314.5	1.3652
1000	7.6172	0.13128	1244.8	1366.4	1.4014
1100	6.5368	0.15298	1313.1	1454.7	1.4600
1200	5.8191	0.17185	1373.0	1532.1	1.5081
1300	5.2904	0.18902	1428.8	1603.8	1.5500
1400	4.8761	0.20508	1482.4	1672.3	1.5879
1500	4.5382	0.22035	1534.8	1738.8	1.6227
1600	4.2545	0.23505	1586.7	1804.3	1.6553
1700	4.0113	0.24930	1638.3	1869.2	1.6861
1800	3.7994	0.26320	1690.1	1933.8	1.7153

Table DE.3JJ $P = 6000$ psia

T, °F	ρ, lbm/ft³	v, ft³/lbm	u, Btu/lbm	h, Btu/lbm	s, Btu/lbm·R
650	43.019	0.023246	640.73	666.56	0.84136
700	39.009	0.025635	708.54	737.03	0.90344
750	33.540	0.029815	789.22	822.34	0.97541
800	25.322	0.039491	897.71	941.58	1.0719
850	17.197	0.058151	1019.2	1083.8	1.1827
900	13.185	0.075844	1104.3	1188.5	1.2612
950	11.099	0.090098	1164.5	1264.6	1.3162
1000	9.7958	0.10208	1212.2	1325.6	1.3587
1100	8.1891	0.12211	1289.3	1425.0	1.4247
1200	7.1885	0.13911	1354.3	1508.8	1.4768
1300	6.4793	0.15434	1413.4	1584.9	1.5213
1400	5.9377	0.16841	1469.3	1656.5	1.5609
1500	5.5039	0.18169	1523.5	1725.4	1.5970
1600	5.1445	0.19438	1576.8	1792.7	1.6305
1700	4.8396	0.20663	1629.6	1859.1	1.6620
1800	4.5761	0.21853	1682.2	1925.0	1.6918

Table DE.4   Compressed Liquid (Water)

Table DE.4A $P = 500$ psia ($T_{sat} = 467.04°F$)

T, °F	ρ, lbm/ft^3	v, ft^3/lbm	u, Btu/lbm	h, Btu/lbm	s, Btu/lbm·R
31.558	62.523	0.015994	−0.43616	1.0447	−0.00089009
50	62.509	0.015998	18.040	19.521	0.036035
100	62.086	0.016107	67.906	69.397	0.12939
150	61.286	0.016317	117.78	119.29	0.21477
200	60.217	0.016607	167.82	169.35	0.29369
250	58.921	0.016972	218.18	219.75	0.36733
300	57.416	0.017417	269.09	270.71	0.43671
350	55.697	0.017954	320.85	322.51	0.50275
400	53.737	0.018609	373.86	375.58	0.56634
450	51.481	0.019425	428.73	430.53	0.62845
467.04	50.628	0.019752	448.01	449.84	0.64949

Table DE.4B $P = 1000$ psia ($T_{sat} = 544.65°F$)

T, °F	ρ, lbm/ft^3	v, ft^3/lbm	u, Btu/lbm	h, Btu/lbm	s, Btu/lbm·R
31.092	62.631	0.015967	−0.87714	2.0795	−0.0017960
50	62.611	0.015972	18.002	20.959	0.035953
100	62.179	0.016083	67.738	70.716	0.12908
150	61.380	0.016292	117.50	120.51	0.21430
200	60.315	0.016580	167.42	170.49	0.29309
250	59.028	0.016941	217.66	220.80	0.36659
300	57.536	0.017380	268.42	271.64	0.43581
350	55.835	0.017910	319.98	323.30	0.50167
400	53.903	0.018552	372.73	376.16	0.56501
450	51.689	0.019347	427.22	430.80	0.62678
500	49.097	0.020368	484.35	488.12	0.68811
544.65	46.297	0.021600	539.02	543.02	0.74400

Table DE.4C $P = 1500$ psia ($T_{sat} = 596.26°F$)

T, °F	ρ, lbm/ft³	v, ft³/lbm	u, Btu/lbm	h, Btu/lbm	s, Btu/lbm·R
30.621	62.738	0.015939	−1.32290	3.1044	−0.0027177
50	62.712	0.015946	17.963	22.392	0.035864
100	62.272	0.016059	67.571	72.032	0.12877
150	61.472	0.016267	117.22	121.74	0.21384
200	60.412	0.016553	167.03	171.63	0.29249
250	59.133	0.016911	217.14	221.84	0.36586
300	57.654	0.017345	267.75	272.57	0.43493
350	55.971	0.017866	319.13	324.09	0.50060
400	54.065	0.018496	371.62	376.76	0.56371
450	51.890	0.019271	425.75	431.11	0.62515
500	49.363	0.020258	482.33	487.96	0.68598
550	46.307	0.021595	542.87	548.86	0.74782
596.26	42.622	0.023462	605.56	612.08	0.80900

Table DE.4D $P = 2000$ psia ($T_{sat} = 635.85°F$)

T, °F	ρ, lbm/ft³	v, ft³/lbm	u, Btu/lbm	h, Btu/lbm	s, Btu/lbm·R
30.144	62.844	0.015912	−1.77360	4.1195	−0.0036552
50	62.812	0.015921	17.922	23.819	0.035769
100	62.365	0.016035	67.407	73.345	0.12847
150	61.564	0.016243	116.95	122.96	0.21338
200	60.508	0.016527	166.65	172.77	0.29190
250	59.237	0.016881	216.64	222.89	0.36513
300	57.770	0.017310	267.10	273.51	0.43405
350	56.105	0.017824	318.29	324.89	0.49955
400	54.223	0.018442	370.55	377.38	0.56244
450	52.086	0.019199	424.34	431.45	0.62357
500	49.618	0.020154	480.40	487.86	0.68393
550	46.671	0.021427	540.02	547.96	0.74495
600	42.888	0.023317	606.17	614.81	0.80953
635.85	39.009	0.025635	662.79	672.28	0.86284

Table DE.4E $P = 3000$ psia ($T_{sat} = 695.41°F$)

T, °F	ρ, lbm/ft³	v, ft³/lbm	u, Btu/lbm	h, Btu/lbm	s, Btu/lbm·R
29.173	63.056	0.015859	−2.68980	6.1201	−0.0055781
50	63.010	0.015870	17.839	26.656	0.035560
100	62.547	0.015988	67.083	75.964	0.12785
150	61.746	0.016195	116.41	125.41	0.21247
200	60.698	0.016475	165.90	175.05	0.29073
250	59.442	0.016823	215.65	225.00	0.36370
300	57.998	0.017242	265.83	275.41	0.43234
350	56.365	0.017741	316.67	326.53	0.49750
400	54.530	0.018338	368.47	378.66	0.55997
450	52.461	0.019062	421.64	432.22	0.62053
500	50.099	0.019960	476.76	487.85	0.68005
550	47.335	0.021126	534.84	546.58	0.73968
600	43.940	0.022759	597.82	610.46	0.80141
650	39.272	0.025463	670.99	685.13	0.87021
695.41	29.121	0.034339	783.94	803.02	0.97391

Table DE.4F $P = 5000$ psia

T, °F	ρ, lbm/ft³	v, ft³/lbm	u, Btu/lbm	h, Btu/lbm	s, Btu/lbm·R
27.159	63.473	0.015755	−4.5835	10.003	−0.0096178
50	63.398	0.015773	17.663	32.267	0.035073
100	62.905	0.015897	66.454	81.172	0.12661
150	62.101	0.016103	115.39	130.29	0.21067
200	61.067	0.016375	164.46	179.63	0.28844
250	59.840	0.016711	213.76	229.23	0.36092
300	58.437	0.017112	263.41	279.26	0.42903
350	56.864	0.017586	313.61	329.90	0.49358
400	55.111	0.018145	364.60	381.40	0.55530
450	53.158	0.018812	416.67	434.09	0.61487
500	50.968	0.019620	470.26	488.42	0.67300
550	48.477	0.020628	525.96	545.06	0.73053
600	45.573	0.021943	584.81	605.12	0.78857
650	42.037	0.023788	648.71	670.73	0.84904
700	37.345	0.026777	722.26	747.05	0.91627

Table DE.5 Vapor Properties: Saturated Solid (Ice)–Vapor

Temp., T °F	Sat. Press., P_{sat} psia	Specific Volume, ft³/lbm		Internal Energy, Btu/lbm			Enthalpy, Btu/lbm			Entropy, Btu/lbm·K		
		Sat. Ice, v_l	Sat. Vapor, $v_g \times 10^{-3}$	Sat. Ice, u_l	Subl., u_{ig}	Sat. Vapor, u_g	Sat. Ice, h_l	Subl., h_{ig}	Sat. Vapor, h_g	Sat. Ice, s_l	Subl., s_{ig}	Sat. Vapor, s_g
32.018	0.0887	0.01747	3.302	−143.34	1164.6	1021.2	−143.34	1218.7	1075.4	−0.292	2.479	2.187
32	0.0886	0.01747	3.305	−143.35	1164.6	1021.2	−143.35	1218.7	1075.4	−0.292	2.479	2.187
30	0.0808	0.01747	3.607	−144.35	1164.9	1020.5	−144.35	1218.9	1074.5	−0.294	2.489	2.195
25	0.0641	0.01746	4.506	−146.84	1165.7	1018.9	−146.84	1219.1	1072.3	−0.299	2.515	2.216
20	0.0505	0.01745	5.655	−149.31	1166.5	1017.2	−149.31	1219.4	1070.1	−0.304	2.542	2.238
15	0.0396	0.01745	7.13	−151.75	1167.3	1015.3	−151.75	1219.7	1067.9	−0.309	2.569	2.260
10	0.0309	0.01744	9.04	−154.17	1168.1	1013.9	−154.17	1219.9	1065.7	−0.314	2.597	2.283
5	0.0240	0.01743	11.52	−156.56	1168.8	1012.2	−156.56	1220.1	1063.5	−0.320	2.626	2.306
0	0.0185	0.01743	14.77	−158.93	1169.5	1010.6	−158.93	1220.2	1061.2	−0.325	2.655	2.330
−5	0.0142	0.01742	19.03	−161.27	1170.2	1008.9	−161.27	1220.3	1059.0	−0.330	2.684	2.354
−10	0.0109	0.01741	24.66	−163.59	1170.9	1007.3	−163.59	1220.4	1056.8	−0.335	2.714	2.379
−15	0.0082	0.01740	32.2	−165.89	1171.5	1005.6	−165.89	1220.5	1054.6	−0.340	2.745	2.405
−20	0.0062	0.01740	42.2	−168.16	1172.1	1003.9	−168.16	1220.6	1052.4	−0.345	2.776	2.431
−25	0.0046	0.01739	55.7	−170.40	1172.7	1002.3	−170.40	1220.6	1050.2	−0.351	2.808	2.457
−30	0.0035	0.01738	74.1	−172.63	1173.2	1000.6	−172.63	1220.6	1048.0	−0.356	2.841	2.485
−35	0.0026	0.01737	99.2	−174.82	1173.8	988.9	−174.82	1220.6	1045.8	−0.361	2.874	2.513
−40	0.0019	0.01737	133.8	−177.00	1174.3	997.3	−177.00	1220.6	1043.6	−0.366	2.908	2.542

Appendix EE
Various Thermodynamic Data

Table EE.1   Critical Constants and Specific Heats for Selected Gases*

Substance	m (lb/lbmol)	T_c (R)	P_c (psia)	v_c (m³/kmol)	Z_c	c_v (kJ/kg·K)	c_p (kJ/kg·K)
Acetylene (C_2H_2)	26.04	556.20	905.04	1.79	0.27	0.33	0.40
Air (equivalent)	28.97	239.40	546.79	1.33	0.28	0.17	0.24
Ammonia (NH_3)	17.04	730.80	1636.03	1.16	0.24	0.40	0.51
Benzene (C_6H_6)	78.11	1011.60	700.53	4.10	0.27	0.16	0.19
n-Butane (C_4H_{10})	58.12	765.36	549.69	4.12	0.27	0.37	0.41
Carbon dioxide (CO_2)	44.01	547.56	1071.83	1.51	0.28	0.16	0.20
Carbon monoxide (CO)	28.01	239.40	507.63	1.49	0.29	0.18	0.25
Refrigerant 134a ($C_2F_4H_2$)	102.03	673.74	588.85	3.20	0.26	0.18	0.20
Ethane (C_2H_6)	30.07	549.72	707.79	2.37	0.29	0.35	0.42
Ethylene (C_2H_4)	28.05	509.40	742.59	2.05	0.28	0.29	0.37
Helium (He)	4.00	9.36	33.36	0.93	0.30	0.75	1.24
Hydrogen (H_2)	2.02	59.76	188.55	1.04	0.30	2.44	3.42
Methane (CH_4)	16.04	343.26	672.98	1.59	0.29	0.41	0.53
Nitrogen (N_2)	28.01	227.16	491.68	1.44	0.29	0.18	0.25
Oxygen (O_2)	32.00	277.92	732.44	1.19	0.29	0.16	0.22
Propane (C_3H_8)	44.09	666.00	616.41	3.20	0.28	0.35	0.40
Sulfur dioxide (SO_2)	64.06	775.80	1141.45	1.99	0.27	0.11	0.14
Water (H_2O)	18.02	1164.78	3199.54	0.89	0.23	0.33	0.44

*Adapted from Wark, K., Jr., and Richards, D. E., *Thermodynamics,* 6th ed., McGraw-Hill, New York, 1999.

Table EE.2   Van der Waals Constants for Selected Gases*

Substance	a [psia·(ft³/lbmol)²]	b (ft³/lbmol)
Acetylene (C_2H_2)	16412	0.82
Air (equivalent)	5053.9	0.58
Ammonia (NH_3)	15716	0.60
Benzene (C_6H_6)	69333	1.89
n-Butane (C_4H_{10})	51358	1.92
Carbon dioxide (CO_2)	13558	0.68
Carbon monoxide (CO)	5444.7	0.63
Refrigerant 134a ($C_2F_4H_2$)	37402	1.53
Ethane (C_2H_6)	20748	1.04
Ethylene (C_2H_4)	16982	0.92
Helium (He)	126.91	0.37
Hydrogen (H_2)	919.23	0.42
Methane (CH_4)	8503.9	0.68
Nitrogen (N_2)	5065.1	0.62
Oxygen (O_2)	5094.9	0.50
Propane (C_3H_8)	34667	1.44
Sulfur dioxide (SO_2)	25445	0.91
Water (H_2O)	20495	0.49

*Adapted from Wark, K., Jr., and Richards, D. E., Thermodynamics, 6th ed., McGraw-Hill, New York, 1999.

Appendix FE
Thermo-Physical Properties of Selected Gases at 1 atm

Table FE.1 Thermo-Physical Properties of Selected Gases (1 atm)

Table FE.1A Ammonia (NH$_3$)

T (R)	ρ (lbm/ft^3)	c_p (Btu/lbm·R)	μ (lbm/ft·s)	ν (ft^2/s)	k (Btu/h·ft·R)	α (ft^2/s)	Pr
400	0.096040	0.24917	0.0000094671	0.000098574	0.011609	0.00013476	0.73146
420	0.091441	0.24908	0.0000098515	0.00010774	0.012117	0.00014778	0.72901
440	0.087264	0.24901	0.000010229	0.00011722	0.012618	0.00016129	0.72674
460	0.083454	0.24896	0.000010600	0.00012701	0.013111	0.00017529	0.72461
480	0.079964	0.24892	0.000010964	0.00013711	0.013596	0.00018974	0.72263
500	0.076754	0.24890	0.000011322	0.00014752	0.014076	0.00020466	0.72077
520	0.073794	0.24889	0.000011675	0.00015821	0.014548	0.00022003	0.71903
540	0.071053	0.24889	0.000012022	0.00016919	0.015014	0.00023584	0.71740
560	0.068510	0.24891	0.000012363	0.00018046	0.015475	0.00025207	0.71588
580	0.066142	0.24895	0.000012699	0.00019200	0.015929	0.00026873	0.71446
600	0.063934	0.24900	0.000013030	0.00020381	0.016378	0.00028579	0.71315
620	0.061868	0.24907	0.000013357	0.00021589	0.016822	0.00030324	0.71194
640	0.059932	0.24916	0.000013679	0.00022824	0.017261	0.00032108	0.71083
660	0.058113	0.24928	0.000013996	0.00024084	0.017695	0.00033930	0.70982
680	0.056402	0.24941	0.000014309	0.00025370	0.018124	0.00035787	0.70891
700	0.054789	0.24957	0.000014618	0.00026681	0.018548	0.00037680	0.70810
720	0.053265	0.24975	0.000014923	0.00028017	0.018968	0.00039606	0.70739
740	0.051825	0.24996	0.000015225	0.00029377	0.019384	0.00041565	0.70678
760	0.050460	0.25019	0.000015522	0.00030762	0.019795	0.00043556	0.70626
780	0.049165	0.25044	0.000015817	0.00032170	0.020203	0.00045577	0.70584
800	0.047935	0.25073	0.000016107	0.00033602	0.020607	0.00047628	0.70552
820	0.046765	0.25103	0.000016395	0.00035057	0.021007	0.00049707	0.70528
840	0.045651	0.25136	0.000016679	0.00036535	0.021404	0.00051813	0.70513
860	0.044589	0.25171	0.000016960	0.00038036	0.021797	0.00053946	0.70508
880	0.043576	0.25209	0.000017238	0.00039559	0.022187	0.00056105	0.70510
900	0.042607	0.25249	0.000017513	0.00041105	0.022574	0.00058288	0.70520
920	0.041681	0.25291	0.000017786	0.00042672	0.022958	0.00060495	0.70538
940	0.040793	0.25336	0.000018056	0.00044261	0.023338	0.00062725	0.70564
960	0.039943	0.25382	0.000018323	0.00045872	0.023716	0.00064978	0.70596
980	0.039128	0.25430	0.000018587	0.00047504	0.024091	0.00067253	0.70635
1000	0.038346	0.25480	0.000018850	0.00049157	0.024463	0.00069548	0.70681

Table FE.1B Carbon Dioxide (CO$_2$)

T (R)	ρ (lbm/ft^3)	c_p (Btu/lbm·R)	μ (lbm/ft·s)	ν (ft^2/s)	k (Btu/h·ft·R)	α (ft^2/s)	Pr
400	0.152720	0.18702	0.0000075092	0.000049170	0.0063872	0.000062120	0.79154
420	0.145150	0.18902	0.0000078839	0.000054317	0.0068142	0.000068993	0.78728
440	0.138320	0.19126	0.000008257	0.000059697	0.0072588	0.000076217	0.78325
460	0.132130	0.19366	0.000008629	0.000065307	0.0077208	0.000083817	0.77916
480	0.126480	0.19615	0.000008998	0.000071145	0.0081988	0.000091801	0.77499
500	0.121310	0.19868	0.000009366	0.000077208	0.0086908	0.00010016	0.77082
520	0.116550	0.20123	0.000009731	0.000083493	0.0091947	0.00010890	0.76671
540	0.112160	0.20378	0.000010094	0.000089996	0.0097085	0.00011799	0.76274
560	0.108090	0.20631	0.000010454	0.000096714	0.010230	0.00012743	0.75897
580	0.104320	0.20881	0.000010812	0.00010365	0.010759	0.00013720	0.75543
600	0.100800	0.21127	0.000011167	0.00011078	0.011292	0.00014729	0.75215
620	0.097508	0.21369	0.000011519	0.00011813	0.011829	0.00015769	0.74913
640	0.094431	0.21607	0.000011868	0.00012568	0.012368	0.00016838	0.74637
660	0.091543	0.21840	0.000012214	0.00013342	0.012910	0.00017937	0.74386
680	0.088829	0.22067	0.000012557	0.00014137	0.013452	0.00019062	0.74160
700	0.086271	0.22291	0.000012898	0.00014950	0.013995	0.00020215	0.73956
720	0.083858	0.22509	0.000013235	0.00015783	0.014537	0.00021393	0.73774
740	0.081578	0.22722	0.000013570	0.00016634	0.015079	0.00022597	0.73611
760	0.079418	0.22931	0.000013901	0.00017503	0.015620	0.00023826	0.73465
780	0.077371	0.23135	0.000014229	0.00018391	0.016160	0.00025078	0.73336
800	0.075427	0.23335	0.000014555	0.00019297	0.016699	0.00026354	0.73222
820	0.073579	0.23530	0.000014878	0.00020220	0.017235	0.00027653	0.73120
840	0.071819	0.23721	0.000015197	0.00021160	0.017770	0.00028975	0.73031
860	0.070142	0.23908	0.000015514	0.00022118	0.018303	0.00030318	0.72952
880	0.068542	0.24091	0.000015828	0.00023092	0.018835	0.00031684	0.72882
900	0.067014	0.24270	0.000016139	0.00024083	0.019364	0.00033071	0.72820
920	0.065552	0.24445	0.000016447	0.00025090	0.019891	0.00034480	0.72767
940	0.064153	0.24617	0.000016752	0.00026113	0.020415	0.00035909	0.72719
960	0.062813	0.24785	0.000017055	0.00027152	0.020938	0.00037360	0.72678
980	0.061527	0.24949	0.000017355	0.00028207	0.021458	0.00038830	0.72641
1000	0.060294	0.25110	0.000017652	0.00029277	0.021976	0.00040321	0.72609
1020	0.059109	0.25268	0.000017946	0.00030362	0.022492	0.00041832	0.72581
1040	0.057969	0.25423	0.000018238	0.00031462	0.023006	0.00043362	0.72556
1060	0.056873	0.25574	0.000018528	0.00032577	0.023517	0.00044912	0.72534
1080	0.055818	0.25723	0.000018814	0.00033706	0.024026	0.00046482	0.72515
1100	0.054801	0.25868	0.000019098	0.00034850	0.024533	0.00048071	0.72498

(*continued*)

Table FE.1B (continued)

T (R)	ρ (lbm/ft³)	c_p (Btu/lbm·R)	μ (lbm/ft·s)	ν (ft²/s)	k (Btu/h·ft·R)	α (ft²/s)	Pr
1120	0.053821	0.26011	0.000019380	0.00036008	0.025037	0.00049679	0.72483
1140	0.052875	0.26151	0.000019659	0.00037181	0.025539	0.00051305	0.72469
1160	0.051962	0.26288	0.000019936	0.00038367	0.026038	0.00052951	0.72457
1180	0.051080	0.26422	0.000020211	0.00039567	0.026536	0.00054615	0.72446
1200	0.050228	0.26553	0.000020483	0.00040780	0.027031	0.00056298	0.72436
1220	0.049403	0.26683	0.000020752	0.00042007	0.027523	0.00057999	0.72426
1240	0.048605	0.26809	0.000021020	0.00043246	0.028014	0.00059718	0.72418
1260	0.047833	0.26933	0.000021285	0.00044499	0.028502	0.00061455	0.72410
1280	0.047085	0.27055	0.000021548	0.00045765	0.028987	0.00063210	0.72402
1300	0.046359	0.27174	0.000021809	0.00047044	0.029471	0.00064983	0.72394
1320	0.045656	0.27290	0.000022068	0.00048335	0.029952	0.00066774	0.72387
1340	0.044974	0.27405	0.000022325	0.00049639	0.030430	0.00068582	0.72379
1360	0.044312	0.27517	0.000022580	0.00050956	0.030907	0.00070408	0.72372
1380	0.043669	0.27627	0.000022832	0.00052284	0.031381	0.00072251	0.72365
1400	0.043045	0.27735	0.000023083	0.00053625	0.031852	0.00074111	0.72358

Table FE.1C Carbon Monoxide (CO)

T (R)	ρ (lbm/ft³)	c_p (Btu/lbm·R)	μ (lbm/ft·s)	ν (ft²/s)	k (Btu/h·ft·R)	α (ft²/s)	Pr
400	0.096064	0.24926	0.0000094416	0.000098285	0.012131	0.00014073	0.69837
420	0.091460	0.24917	0.0000098161	0.00010733	0.012622	0.00015385	0.69762
440	0.087279	0.24911	0.000010183	0.00011667	0.013103	0.00016740	0.69695
460	0.083465	0.24907	0.000010542	0.00012631	0.013574	0.00018138	0.69637
480	0.079972	0.24904	0.000010895	0.00013623	0.014037	0.00019578	0.69585
500	0.076761	0.24904	0.000011241	0.00014644	0.014493	0.00021059	0.69538
520	0.073798	0.24905	0.000011581	0.00015693	0.014941	0.00022581	0.69496
540	0.071056	0.24909	0.000011915	0.00016768	0.015382	0.00024142	0.69459
560	0.068512	0.24914	0.000012244	0.00017871	0.015818	0.00025741	0.69425
580	0.066143	0.24923	0.000012567	0.00018999	0.016248	0.00027378	0.69395
600	0.063933	0.24933	0.000012885	0.00020153	0.016672	0.00029053	0.69367
620	0.061867	0.24947	0.000013198	0.00021332	0.017093	0.00030764	0.69343
640	0.059930	0.24963	0.000013506	0.00022537	0.017509	0.00032510	0.69321
660	0.058111	0.24982	0.000013810	0.00023765	0.017922	0.00034292	0.69302
680	0.056400	0.25004	0.000014110	0.00025017	0.018331	0.00036108	0.69285
700	0.054786	0.25029	0.000014405	0.00026293	0.018737	0.00037958	0.69270
720	0.053262	0.25056	0.000014696	0.00027593	0.019141	0.00039841	0.69256
740	0.051821	0.25087	0.000014984	0.00028915	0.019543	0.00041757	0.69245
760	0.050456	0.25121	0.000015268	0.00030260	0.019943	0.00043705	0.69235
780	0.049161	0.25158	0.000015548	0.00031627	0.020341	0.00045685	0.69227
800	0.047931	0.25197	0.000015825	0.00033016	0.020738	0.00047696	0.69221
820	0.046762	0.25239	0.000016099	0.00034427	0.021133	0.00049739	0.69216
840	0.045648	0.25284	0.000016369	0.00035859	0.021528	0.00051811	0.69212
860	0.044585	0.25332	0.000016636	0.00037313	0.021921	0.00053914	0.69209
880	0.043572	0.25382	0.000016901	0.00038788	0.022314	0.00056047	0.69207

Table FE.1D   Helium (He)

T (R)	ρ (lbm/ft³)	c_p (Btu/lbm·R)	μ (lbm/ft·s)	ν (ft²/s)	k (Btu/h·ft·R)	α (ft²/s)	Pr
150.00	0.036478	1.2417	0.0000058611	0.00016067	0.037735	0.00023141	0.69432
200.00	0.027370	1.2414	0.0000068967	0.00025198	0.045745	0.00037399	0.67377
250.00	0.021902	1.2413	0.0000079843	0.00036455	0.053192	0.00054350	0.67075
300.00	0.018255	1.2412	0.0000090085	0.00049349	0.060218	0.00073824	0.66846
350.00	0.015649	1.2412	0.0000099848	0.00063804	0.066911	0.00095690	0.66678
400.00	0.013694	1.2412	0.000010923	0.00079760	0.073331	0.0011984	0.66554
450.00	0.012174	1.2412	0.000011828	0.00097163	0.079520	0.0014619	0.66463
500.00	0.010957	1.2412	0.000012707	0.0011597	0.085512	0.0017466	0.66395
550.00	0.0099615	1.2412	0.000013561	0.0013614	0.091332	0.0020520	0.66345
600.00	0.0091318	1.2412	0.000014395	0.0015764	0.096998	0.0023773	0.66311
650.00	0.0084296	1.2412	0.000015211	0.0018044	0.10253	0.0027221	0.66287
700.00	0.0078277	1.2412	0.000016009	0.0020451	0.10793	0.0030860	0.66271
750.00	0.0073061	1.2412	0.000016792	0.0022983	0.11323	0.0034685	0.66261
800.00	0.0068496	1.2412	0.000017560	0.0025637	0.11842	0.0038693	0.66256
850.00	0.0064468	1.2412	0.000018316	0.0028410	0.12352	0.0042881	0.66255
900.00	0.0060888	1.2412	0.000019059	0.0031302	0.12853	0.0047244	0.66257
950.00	0.0057684	1.2412	0.000019791	0.0034310	0.13346	0.0051780	0.66261
1000.0	0.0054801	1.2412	0.000020513	0.0037432	0.13831	0.0056487	0.66267
1050.0	0.0052192	1.2412	0.000021225	0.0040667	0.14309	0.0061361	0.66276
1100.0	0.0049820	1.2412	0.000021928	0.0044014	0.14781	0.0066401	0.66285
1150.0	0.0047654	1.2412	0.000022622	0.0047470	0.15246	0.0071604	0.66296
1200.0	0.0045669	1.2412	0.000023307	0.0051035	0.15706	0.0076967	0.66307
1250.0	0.0043843	1.2412	0.000023985	0.0054707	0.16160	0.0082490	0.66319
1300.0	0.0042157	1.2412	0.000024655	0.0058485	0.16608	0.0088170	0.66332
1350.0	0.0040596	1.2412	0.000025319	0.0062368	0.17052	0.0094005	0.66345
1400.0	0.0039146	1.2412	0.000025975	0.0066355	0.17490	0.0099994	0.66359
1450.0	0.0037797	1.2412	0.000026626	0.0070445	0.17924	0.010613	0.66373
1500.0	0.0036537	1.2412	0.000027270	0.0074636	0.18354	0.011243	0.66387
1550.0	0.0035358	1.2412	0.000027908	0.0078928	0.18779	0.011887	0.66401
1600.0	0.0034254	1.2412	0.000028540	0.0083320	0.19201	0.012545	0.66416
1650.0	0.0033216	1.2412	0.000029167	0.0087812	0.19618	0.013219	0.66430
1700.0	0.0032239	1.2412	0.000029789	0.0092401	0.20032	0.013906	0.66445
1750.0	0.0031318	1.2412	0.000030406	0.0097088	0.20442	0.014609	0.66460
1800.0	0.0030448	1.2412	0.000031018	0.010187	0.20849	0.015325	0.66474

Table FE.1E Hydrogen (H$_2$)

T (R)	ρ (lbm/ft³)	c$_p$ (Btu/lbm·R)	μ (lbm/ft·s)	ν (ft²/s)	k (Btu/h·ft·R)	α (ft²/s)	Pr
180.00	0.015340	2.6839	0.0000028153	0.00018353	0.039511	0.00026657	0.68847
200.00	0.013801	2.7598	0.0000030327	0.00021974	0.043766	0.00031918	0.68844
220.00	0.012544	2.8365	0.0000032418	0.00025843	0.047959	0.00037441	0.69025
240.00	0.011497	2.9108	0.0000034440	0.00029956	0.052192	0.00043322	0.69147
260.00	0.010612	2.9802	0.0000036404	0.00034306	0.056338	0.00049485	0.69326
280.00	0.0098531	3.0437	0.0000038317	0.00038888	0.060479	0.00056018	0.69420
300.00	0.0091959	3.1008	0.0000040184	0.00043698	0.064653	0.00062982	0.69382
320.00	0.0086210	3.1516	0.0000042011	0.00048731	0.068720	0.00070258	0.69361
340.00	0.0081138	3.1964	0.0000043802	0.00053984	0.072706	0.00077873	0.69323
360.00	0.0076630	3.2356	0.0000045558	0.00059452	0.076572	0.00085785	0.69303
380.00	0.0072598	3.2698	0.0000047283	0.00065131	0.080356	0.00094030	0.69266
400.00	0.0068968	3.2997	0.0000048979	0.00071017	0.084019	0.0010256	0.69248
420.00	0.0065685	3.3255	0.0000050648	0.00077107	0.087618	0.0011142	0.69203
440.00	0.0062700	3.3479	0.0000052291	0.00083398	0.091128	0.0012059	0.69159
460.00	0.0059975	3.3673	0.0000053909	0.00089886	0.094536	0.0013003	0.69126
480.00	0.0057477	3.3839	0.0000055504	0.00096568	0.097879	0.0013979	0.69081
500.00	0.0055178	3.3982	0.0000057077	0.0010344	0.10111	0.0014979	0.69057
520.00	0.0053057	3.4103	0.0000058630	0.0011050	0.10431	0.0016014	0.69004
540.00	0.0051093	3.4206	0.0000060162	0.0011775	0.10740	0.0017071	0.68978
560.00	0.0049269	3.4292	0.0000061676	0.0012518	0.11047	0.0018163	0.68921
580.00	0.0047570	3.4364	0.0000063171	0.0013279	0.11350	0.0019286	0.68855
600.00	0.0045985	3.4423	0.0000064649	0.0014059	0.11676	0.0020490	0.68612
620.00	0.0044503	3.4471	0.0000066110	0.0014855	0.11999	0.0021727	0.68372
640.00	0.0043112	3.4509	0.0000067555	0.0015670	0.12319	0.0023000	0.68128
660.00	0.0041807	3.4539	0.0000068985	0.0016501	0.12624	0.0024286	0.67945
680.00	0.0040578	3.4563	0.0000070401	0.0017350	0.12932	0.0025613	0.67737
700.00	0.0039419	3.4581	0.0000071802	0.0018215	0.13233	0.0026966	0.67548

Table FE.1F Nitrogen (N₂)

T (R)	ρ (lbm/ft³)	c_p (Btu/lbm·R)	μ (lbm/ft·s)	v (ft²/s)	k (Btu/h·ft·R)	α (ft²/s)	Pr
200.00	0.19464	0.25388	0.000051604	0.000026512	0.0060360	0.000033930	0.78139
300.00	0.12841	0.25004	0.0000074278	0.000057846	0.0089427	0.000077369	0.74767
400.00	0.096040	0.24917	0.0000094671	0.000098574	0.011609	0.00013476	0.73146
500.00	0.076754	0.24890	0.000011322	0.00014752	0.014076	0.00020466	0.72077
600.00	0.063934	0.24900	0.000013030	0.00020381	0.016378	0.00028579	0.71315
700.00	0.054789	0.24957	0.000014618	0.00026681	0.018548	0.00037680	0.70810
800.00	0.047935	0.25073	0.000016107	0.00033602	0.020607	0.00047628	0.70552
900.00	0.042607	0.25249	0.000017513	0.00041105	0.022574	0.00058288	0.70520
1000.0	0.038346	0.25480	0.000018850	0.00049157	0.024463	0.00069548	0.70681
1100.0	0.034859	0.25754	0.000020126	0.00057734	0.026285	0.00081331	0.70987
1200.0	0.031955	0.26055	0.000021350	0.00066814	0.028051	0.00093586	0.71393
1300.0	0.029497	0.26373	0.000022529	0.00076379	0.029766	0.0010629	0.71858
1400.0	0.027390	0.26695	0.000023669	0.00086414	0.031439	0.0011944	0.72350
1500.0	0.025565	0.27014	0.000024774	0.00096907	0.033074	0.0013303	0.72844
1600.0	0.023967	0.27323	0.000025848	0.0010785	0.034675	0.0014708	0.73324
1700.0	0.022558	0.27620	0.000026894	0.0011922	0.036246	0.0016160	0.73776
1800.0	0.021305	0.27901	0.000027916	0.0013103	0.037792	0.0017660	0.74196
1900.0	0.020184	0.28166	0.000028915	0.0014326	0.039313	0.0019209	0.74580
2000.0	0.019175	0.28414	0.000029895	0.0015591	0.040814	0.0020809	0.74925
2100.0	0.018262	0.28646	0.000030857	0.0016897	0.042296	0.0022459	0.75234
2200.0	0.017432	0.28862	0.000031802	0.0018244	0.043762	0.0024162	0.75507
2300.0	0.016674	0.29062	0.000032732	0.0019631	0.045212	0.0025917	0.75745
2400.0	0.015980	0.29248	0.000033649	0.0021058	0.046648	0.0027725	0.75953
2500.0	0.015341	0.29421	0.000034554	0.0022525	0.048072	0.0029587	0.76131
2600.0	0.014751	0.29581	0.000035448	0.0024031	0.049485	0.0031503	0.76283
2700.0	0.014204	0.29730	0.000036331	0.0025577	0.050888	0.0033473	0.76411
2800.0	0.013697	0.29869	0.000037205	0.0027162	0.052282	0.0035498	0.76518
2900.0	0.013225	0.29997	0.000038070	0.0028786	0.053668	0.0037578	0.76605
3000.0	0.012784	0.30117	0.000038927	0.0030449	0.055046	0.0039713	0.76674
3100.0	0.012372	0.30229	0.000039777	0.0032151	0.056418	0.0041903	0.76728
3200.0	0.011985	0.30334	0.000040621	0.0033892	0.057784	0.0044149	0.76767
3300.0	0.011622	0.30432	0.000041458	0.0035671	0.059144	0.0046450	0.76795
3400.0	0.011281	0.30523	0.000042290	0.0037490	0.060499	0.0048808	0.76811
3500.0	0.010958	0.30609	0.000043117	0.0039346	0.061851	0.0051221	0.76817

Table FE.1G   Oxygen (O₂)

T (R)	ρ (lbm/ft³)	c_p (Btu/lbm·R)	μ (lbm/ft·s)	ν (ft²/s)	k (Btu/h·ft·R)	α (ft²/s)	Pr
200.00	0.22303	0.22263	0.0000057500	0.000025781	0.0058675	0.000032826	0.78540
250.00	0.17691	0.22043	0.0000071165	0.000040227	0.0073766	0.000052546	0.76556
300.00	0.14685	0.21921	0.0000084142	0.000057297	0.0088432	0.000076305	0.75088
350.00	0.12562	0.21866	0.0000096497	0.000076818	0.010266	0.00010382	0.73993
400.00	0.10978	0.21851	0.000010829	0.000098640	0.011647	0.00013486	0.73142
450.00	0.097513	0.21869	0.000011958	0.00012263	0.012988	0.00016918	0.72485
500.00	0.087719	0.21921	0.000013041	0.00014867	0.014294	0.00020648	0.72001
550.00	0.079719	0.22005	0.000014083	0.00017666	0.015565	0.00024647	0.71675
600.00	0.073058	0.22122	0.000015087	0.00020651	0.016807	0.00028886	0.71491
650.00	0.067427	0.22266	0.000016057	0.00023814	0.018020	0.00033340	0.71429
700.00	0.062604	0.22434	0.000016996	0.00027149	0.019207	0.00037988	0.71467
750.00	0.058425	0.22620	0.000017906	0.00030648	0.020370	0.00042815	0.71583
800.00	0.054770	0.22818	0.000018790	0.00034307	0.021510	0.00047810	0.71756
850.00	0.051546	0.23025	0.000019649	0.00038120	0.022631	0.00052967	0.71968
900.00	0.048680	0.23235	0.000020486	0.00042082	0.023732	0.00058282	0.72205
950.00	0.046117	0.23446	0.000021302	0.00046191	0.024815	0.00063752	0.72454
1000.0	0.043810	0.23654	0.000022098	0.00050441	0.025882	0.00069377	0.72706
1050.0	0.041723	0.23859	0.000022876	0.00054829	0.026934	0.00075157	0.72953
1100.0	0.039826	0.24057	0.000023638	0.00059353	0.027971	0.00081094	0.73190
1150.0	0.038094	0.24248	0.000024384	0.00064009	0.028994	0.00087190	0.73414
1200.0	0.036507	0.24432	0.000025115	0.00068795	0.030005	0.00093444	0.73621
1250.0	0.035046	0.24608	0.000025832	0.00073708	0.031004	0.00099859	0.73812
1300.0	0.033698	0.24776	0.000026536	0.00078746	0.031991	0.0010644	0.73984
1350.0	0.032450	0.24936	0.000027228	0.00083907	0.032968	0.0011318	0.74139
1400.0	0.031291	0.25087	0.000027908	0.00089189	0.033935	0.0012008	0.74275
1450.0	0.030212	0.25231	0.000028578	0.00094590	0.034892	0.0012715	0.74394
1500.0	0.029205	0.25367	0.000029237	0.0010011	0.035841	0.0013438	0.74497
1550.0	0.028263	0.25497	0.000029887	0.0010574	0.036780	0.0014178	0.74584
1600.0	0.027380	0.25619	0.000030527	0.0011149	0.037712	0.0014934	0.74656
1650.0	0.026551	0.25735	0.000031158	0.0011735	0.038636	0.0015707	0.74715
1700.0	0.025770	0.25845	0.000031781	0.0012333	0.039553	0.0016497	0.74760
1750.0	0.025034	0.25949	0.000032397	0.0012941	0.040463	0.0017303	0.74794

Table FE.1H   Water Vapor (H$_2$O)

T (R)	ρ (lbm/ft^3)	c_p (Btu/lbm·R)	μ (lbm/ft·s)	v (ft^2/s)	k (Btu/h·ft·R)	α (ft^2/s)	Pr
671.62	0.037310	0.49712	0.0000082439	0.00022096	0.014508	0.00021728	1.0169
700.00	0.035692	0.48506	0.0000086417	0.00024212	0.015144	0.00024298	0.99647
800.00	0.031049	0.47248	0.000010096	0.00032515	0.017740	0.00033591	0.96794
900.00	0.027526	0.47362	0.000011604	0.00042157	0.020736	0.00044183	0.95416
1000.0	0.024738	0.47893	0.000013143	0.00053129	0.024019	0.00056312	0.94348
1100.0	0.022470	0.48593	0.000014696	0.00065400	0.027527	0.00070027	0.93392
1200.0	0.020587	0.49377	0.000016249	0.00078930	0.031220	0.00085313	0.92517
1300.0	0.018996	0.50215	0.0000017795	0.00093678	0.035072	0.0010213	0.91722
1400.0	0.017635	0.51090	0.000019328	0.0010960	0.039063	0.0012044	0.91004
1500.0	0.016456	0.51992	0.000020844	0.0012666	0.043178	0.0014018	0.90356
1600.0	0.015426	0.52913	0.000022339	0.0014482	0.047401	0.0016132	0.89772
1700.0	0.014517	0.53846	0.000023812	0.0016403	0.051723	0.0018380	0.89241
1800.0	0.013709	0.54783	0.000025261	0.0018426	0.056133	0.0020761	0.88752

Appendix GE
Thermo-Physical Properties of Selected Liquids

Table GE.1 Thermo-Physical Properties of Saturated Water

Temperature (R)	Pressure (psia)	Liquid Density (lbm/ft³)	Vapor Density (lbm/ft³)	Liqid c_p (Btm/lbm·R)	Vapor c_p (Btm/lbm·R)	Liquid Viscosity (lbm/ft·s)	Vapor Viscosity (lbm/ft·s)	Liquid Therm. Cond. (Btu/h·ft·R)	Vapor Therm. Cond. (Btu/h·ft·R)	Liquid Prandtl	Vapor Prandtl	Surface Tension (lbm/ft)	Liquid Expansion Coef. β (1/R)	Vapor Expansion Coef. β (1/R)	T (R)
500	0.0083906	62.423	0.00041430	1.0053	0.45149	0.0010323	0.0000062672	0.32945	0.010013	11.341	1.0173	0.0051388	0.000055253	0.0020128	500
510	0.012271	62.405	0.00059414	1.0027	0.45289	0.00087297	0.0000063608	0.33554	0.010194	9.3912	1.0173	0.0050839	0.000050080	0.0019760	510
520	0.017653	62.362	0.00083847	1.0010	0.45436	0.00074972	0.0000064587	0.34153	0.010385	7.9104	1.0172	0.0050278	0.000088515	0.0019409	520
530	0.025005	62.296	0.0011656	0.99988	0.45588	0.00065227	0.0000065604	0.34732	0.010586	6.7601	1.0171	0.0049706	0.00012234	0.0019075	530
540	0.034906	62.210	0.0015975	0.99926	0.45748	0.00057376	0.0000066656	0.35285	0.010796	5.8495	1.0168	0.0049120	0.00015262	0.0018758	540
550	0.048061	62.107	0.0021605	0.99897	0.45915	0.00050947	0.0000067739	0.35806	0.011016	5.1171	1.0164	0.0048524	0.00018009	0.0018457	550
560	0.065323	61.987	0.0028854	0.99893	0.46094	0.00045611	0.0000068850	0.36290	0.011245	4.5198	1.0160	0.0047915	0.00020534	0.0018173	560
570	0.087705	61.853	0.0038081	0.99907	0.46285	0.00041129	0.0000069985	0.36737	0.011484	4.0266	1.0155	0.0047295	0.00022879	0.0017904	570
580	0.11640	61.705	0.0049701	0.99935	0.46492	0.00037325	0.0000071143	0.37144	0.011732	3.6151	1.0149	0.0046663	0.00025078	0.0017652	580
590	0.15281	61.544	0.0064186	0.99977	0.46719	0.00034066	0.0000072321	0.37513	0.011990	3.2684	1.0144	0.0046020	0.00027156	0.0017415	590
600	0.19853	61.371	0.0082072	1.0003	0.46967	0.00031253	0.0000073516	0.37843	0.012259	2.9739	1.0140	0.0045366	0.00029136	0.0017195	600
610	0.25542	61.187	0.010396	1.0009	0.47242	0.00028806	0.0000074728	0.38137	0.012538	2.7217	1.0137	0.0044701	0.00031034	0.0016992	610
620	0.32558	60.992	0.013052	1.0017	0.47545	0.00026665	0.0000075953	0.38396	0.012828	2.5043	1.0135	0.0044025	0.00032867	0.0016805	620
630	0.41137	60.786	0.016249	1.0025	0.47881	0.00024780	0.0000077191	0.38622	0.013129	2.3156	1.0135	0.0043338	0.00034646	0.0016636	630
640	0.51543	60.571	0.020070	1.0035	0.48253	0.00023112	0.0000078439	0.38818	0.013441	2.1510	1.0137	0.0042642	0.00036382	0.0016484	640
650	0.64071	60.346	0.024602	1.0046	0.48665	0.00021630	0.0000079696	0.38986	0.013765	2.0066	1.0143	0.0041934	0.00038085	0.0016350	650
660	0.79045	60.112	0.029944	1.0059	0.49121	0.00020306	0.0000080961	0.39128	0.014102	1.8793	1.0152	0.0041217	0.00039763	0.0016235	660
670	0.96823	59.869	0.036201	1.0073	0.49625	0.00019118	0.0000082232	0.39246	0.014451	1.7666	1.0166	0.0040490	0.00041425	0.0016138	670
671.67	1.0009	59.828	0.037343	1.0076	0.49714	0.00018932	0.0000082445	0.39263	0.014510	1.7490	1.0169	0.0040367	0.00041701	0.0016124	671.67
680	1.1779	59.617	0.043486	1.0089	0.50181	0.00018050	0.0000083509	0.39341	0.014812	1.6664	1.0185	0.0039753	0.00043077	0.0016061	680
690	1.4238	59.357	0.051922	1.0107	0.50795	0.00017085	0.0000084790	0.39416	0.015187	1.5770	1.0210	0.0039007	0.00044726	0.0016005	690
700	1.7103	59.088	0.061639	1.0126	0.51469	0.00016210	0.0000086075	0.39471	0.015574	1.4971	1.0240	0.0038251	0.00046379	0.0015969	700
710	2.0425	58.811	0.072779	1.0147	0.52209	0.00015414	0.0000087361	0.39507	0.015975	1.4253	1.0278	0.0037486	0.00048042	0.0015955	710
720	2.4256	58.525	0.085489	1.0171	0.53019	0.00014689	0.0000088649	0.39526	0.016390	1.3607	1.0324	0.0036713	0.00049722	0.0015964	720
730	2.8651	58.232	0.099930	1.0197	0.53901	0.00014026	0.0000089938	0.39529	0.016818	1.3025	1.0377	0.0035930	0.00051424	0.0015995	730
740	3.3670	57.930	0.11627	1.0225	0.54861	0.00013418	0.0000091228	0.39516	0.017261	1.2499	1.0438	0.0035140	0.00053154	0.0016051	740
750	3.9377	57.619	0.13469	1.0255	0.55900	0.00012858	0.0000092517	0.39487	0.017717	1.2022	1.0509	0.0034341	0.00054920	0.0016130	750
760	4.5837	57.301	0.15538	1.0288	0.57020	0.00012343	0.0000093806	0.39443	0.018188	1.1590	1.0587	0.0033534	0.00056727	0.0016235	760
770	5.3120	56.974	0.17854	1.0324	0.58225	0.00011867	0.0000095094	0.39384	0.018673	1.1199	1.0675	0.0032719	0.00058582	0.0016365	770
780	6.1300	56.639	0.20439	1.0363	0.59515	0.00011426	0.0000096381	0.39311	0.019172	1.0843	1.0771	0.0031898	0.00060494	0.0016522	780
790	7.0453	56.295	0.23315	1.0405	0.60892	0.00011016	0.0000097668	0.39223	0.019686	1.0520	1.0876	0.0031068	0.00062468	0.0016706	790
800	8.0659	55.943	0.26507	1.0450	0.62358	0.00010635	0.0000098953	0.39120	0.020215	1.0227	1.0989	0.0030233	0.00064515	0.0016919	800
810	9.2001	55.582	0.30040	1.0499	0.63916	0.00010280	0.000010024	0.39003	0.020759	0.99615	1.1111	0.0029390	0.00066641	0.0017161	810
820	10.457	55.212	0.33942	1.0551	0.65566	0.000099476	0.000010152	0.38871	0.021318	0.97208	1.1241	0.0028541	0.00068858	0.0017433	820
830	11.844	54.833	0.38242	1.0608	0.67313	0.000096366	0.000010281	0.38724	0.021892	0.95033	1.1380	0.0027686	0.00071176	0.0017738	830

840	0.0018078	0.00073606	0.0026826	1.1528	0.93072	0.022483	0.38561	0.000010409	0.000093446	0.69161	1.0669	0.42970	54.445	13.372	840
850	0.0018453	0.00076161	0.0025960	1.1685	0.91313	0.023089	0.38383	0.000010538	0.000090700	0.71115	1.0734	0.48160	54.046	15.050	850
860	0.0018868	0.00078856	0.0025090	1.1851	0.89744	0.023713	0.38188	0.000010667	0.000088112	0.73181	1.0804	0.53849	53.638	16.888	860
870	0.0019325	0.00081707	0.0024214	1.2028	0.88355	0.024356	0.37977	0.000010797	0.000085666	0.75368	1.0880	0.60073	53.219	18.896	870
880	0.0019828	0.00084732	0.0023335	1.2216	0.87137	0.025017	0.37749	0.000010927	0.000083351	0.77687	1.0962	0.66876	52.790	21.085	880
890	0.0020381	0.00087952	0.0022452	1.2416	0.86085	0.025699	0.37502	0.000011058	0.000081153	0.80149	1.1050	0.74301	52.349	23.465	890
900	0.0020989	0.00091391	0.0021565	1.2628	0.85195	0.026403	0.37237	0.000011190	0.000079063	0.82770	1.1146	0.82398	51.897	26.047	900
910	0.0021659	0.00095076	0.0020676	1.2856	0.84463	0.027132	0.36953	0.000011323	0.000077071	0.85567	1.1249	0.91220	51.433	28.843	910
920	0.0022398	0.00099039	0.0019784	1.3098	0.83889	0.027888	0.36649	0.000011458	0.000075167	0.88561	1.1362	1.0083	50.955	31.864	920
930	0.0023213	0.0010332	0.0018890	1.3359	0.83473	0.028675	0.36323	0.000011594	0.000073343	0.91778	1.1484	1.1128	50.464	35.122	930
940	0.0024115	0.0010796	0.0017995	1.3638	0.83219	0.029498	0.35976	0.000011732	0.000071591	0.95248	1.1617	1.2265	49.959	38.631	940
950	0.0025117	0.0011300	0.0017099	1.3939	0.83130	0.030361	0.35606	0.000011873	0.000069903	0.99005	1.1762	1.3503	49.439	42.401	950
960	0.0026231	0.0011853	0.0016202	1.4263	0.83215	0.031271	0.35211	0.000012018	0.000068274	1.0309	1.1921	1.4849	48.902	46.447	960
970	0.0027476	0.0012460	0.0015307	1.4613	0.83482	0.032236	0.34792	0.000012165	0.000066695	1.0756	1.2097	1.6314	48.348	50.782	970
980	0.0028873	0.0013131	0.0014412	1.4992	0.83944	0.033267	0.34348	0.000012317	0.000065161	1.1248	1.2291	1.7910	47.776	55.418	980
990	0.0030447	0.0013878	0.0013519	1.5402	0.84616	0.034376	0.33878	0.000012474	0.000063666	1.1791	1.2507	1.9649	47.184	60.372	990
1000	0.0032232	0.0014714	0.0012629	1.5848	0.85519	0.035579	0.33383	0.000012637	0.000062204	1.2395	1.2749	2.1546	46.569	65.657	1000
1010	0.0034268	0.0015657	0.0011743	1.6334	0.86678	0.036896	0.32862	0.000012807	0.000060768	1.3072	1.3020	2.3620	45.932	71.288	1010
1020	0.0036607	0.0016731	0.0010861	1.6865	0.88124	0.038352	0.32316	0.000012985	0.000059353	1.3837	1.3328	2.5890	45.268	77.281	1020
1030	0.0039317	0.0017964	0.00099851	1.7446	0.89898	0.039979	0.31748	0.000013173	0.000057952	1.4708	1.3680	2.8382	44.576	83.654	1030
1040	0.0042488	0.0019397	0.00091160	1.8086	0.92050	0.041817	0.31158	0.000013372	0.000056560	1.5711	1.4086	3.1125	43.852	90.422	1040
1050	0.0046240	0.0021082	0.00082552	1.8796	0.94652	0.043921	0.30550	0.000013586	0.000055168	1.6879	1.4560	3.4154	43.092	97.604	1050
1060	0.0050738	0.0023093	0.00074042	1.9591	0.97798	0.046359	0.29927	0.000013817	0.000053770	1.8259	1.5120	3.7515	42.291	105.22	1060
1070	0.0056220	0.0025535	0.00065648	2.0491	1.0163	0.049225	0.29292	0.000014069	0.000052256	1.9915	1.5793	4.1264	41.443	113.29	1070
1080	0.0063030	0.0028564	0.00057391	2.1528	1.0634	0.052644	0.28646	0.000014347	0.000050917	2.1943	1.6619	4.5474	40.541	121.83	1080
1090	0.0071700	0.0032426	0.00049296	2.2751	1.1226	0.056793	0.27993	0.000014658	0.000049440	2.4486	1.7657	5.0241	39.575	130.88	1090
1100	0.0083084	0.0037539	0.00041392	2.4242	1.1995	0.061926	0.27333	0.000015012	0.000047908	2.7778	1.9009	5.5697	38.530	140.45	1100
1110	0.0098661	0.0044671	0.00033718	2.6142	1.3042	0.068428	0.26663	0.000015424	0.000046297	3.2216	2.0863	6.2031	37.385	150.57	1110
1120	0.012124	0.0055396	0.00026320	2.8719	1.4569	0.076921	0.25981	0.000015915	0.000044571	3.8558	2.3590	6.9528	36.108	161.28	1120
1130	0.015694	0.0073323	0.00019266	3.2541	1.7020	0.088528	0.25286	0.000016522	0.000042663	4.8434	2.8022	7.8653	34.639	172.62	1130
1140	0.022209	0.010783	0.00012653	3.9060	2.1463	0.10565	0.24592	0.000017316	0.000040460	6.6196	3.6236	9.0271	32.872	184.63	1140
1150	0.038052	0.018932	0.000066460	5.3425	3.1066	0.13530	0.24024	0.000018466	0.000037757	10.873	5.4907	10.640	30.606	197.36	1150
1160	0.13140	0.069836	0.000016166	11.861	7.9724	0.22216	0.25709	0.000020665	0.000033800	35.418	16.844	13.517	27.112	210.93	1160

Table GE.2   Thermo-Physical Properties of Saturated Refrigerant-134a

T (R)	ρ (lbm/ft^3)	c_p (Btm/lbm·R)	μ (lbm/ft·s)	ν (ft^2/s)	k (Btu/hr·ft·R)	α (ft^2/s)	Pr	β (1/R)
320	98.020	0.28344	0.0010698	0.000010914	0.081199	0.00000081183	13.443	0.00093918
340	96.169	0.28536	0.00076562	0.0000079612	0.077450	0.00000078396	10.155	0.00096800
360	94.295	0.28818	0.00058280	0.0000061806	0.073859	0.00000075499	8.1863	0.0010011
380	92.392	0.29162	0.00046169	0.0000049971	0.070409	0.00000072588	6.8841	0.0010394
400	90.454	0.29555	0.00037618	0.0000041588	0.067090	0.00000069710	5.9659	0.0010842
420	88.471	0.29994	0.00031292	0.0000035370	0.063890	0.00000066879	5.2886	0.0011372
440	86.435	0.30482	0.0002436	0.0000030585	0.060795	0.00000064096	4.7718	0.0012003
460	84.332	0.31028	0.00022594	0.0000026792	0.057793	0.00000061351	4.3670	0.0012767
480	82.149	0.31648	0.00019475	0.0000023708	0.054872	0.00000058628	4.0437	0.0013704
500	79.866	0.32363	0.00016886	0.0000021143	0.052019	0.00000055904	3.7820	0.0014876
520	77.458	0.33208	0.00014692	0.0000018967	0.049217	0.00000053150	3.5687	0.0016376
540	74.893	0.34236	0.00012798	0.0000017088	0.046451	0.00000050324	3.3957	0.0018351
560	72.124	0.35535	0.00011134	0.0000015437	0.043703	0.00000047367	3.2590	0.0021052
580	69.084	0.37260	0.000096446	0.0000013961	0.040950	0.00000044190	3.1592	0.0024940
600	65.668	0.39729	0.000082849	0.0000012616	0.038166	0.00000040637	3.1047	0.0030970
620	61.692	0.43705	0.000070113	0.0000011365	0.035322	0.00000036389	3.1231	0.0041552
640	56.764	0.51761	0.000057656	0.0000010157	0.032407	0.00000030638	3.3152	0.0065166
660	49.604	0.81814	0.000044059	0.00000088822	0.029858	0.00000020437	4.3461	0.016461

Appendix HE
Thermo-Physical Properties of Hydrocarbon Fuels

Table HE.1 Selected Properties of Hydrocarbon Fuels: Enthalpy of Formation,[a] Gibbs Function of Formation,[a] Entropy,[a] and Higher and Lower Heating Values All at 298.15 K and 1 atm; Boiling Points[b] and Latent Heat of Vaporization[c] at 1 atm; Constant-Pressure Adiabatic Flame Temperature at 1 atm;[d] Liquid Density[c]

Formula	Fuel	Molecular Weight (lb/lbmol)	$\bar{h}_f^\circ$ (Btu/lbmol)	$\Delta \bar{g}_f^\circ$ (Btu/lbmol)	$\bar{s}^\circ$ (Btu/lbmol·R)	HHV* (Btu/lbm)	LHV* (Btu/lbm)	Boiling Pt. (°F)	h_{fg} (Btu/lbm)	$T_{ad}^\dagger$ (R)	$\rho_{liq}^\ddagger$ (lbm/ft³)
CH_4	Methane	16.043	−30873.6	−20956.5	42.6761	23,873	21,503	−263.2	219	4007	18.7
C_2H_2	Acetylene	26.038	93551.1	86311.2	46.0297	21,463	20,733	−119.2	—	4570	—
C_2H_4	Ethene	28.054	21570.8	28106.4	50.3865	21,631	20,276	−154.7	—	4264	—
C_2H_6	Ethane	30.069	−34931.7	−13568.0	52.6018	22,313	20,417	−127.5	210	4066	23.1
C_3H_6	Propene	42.080	8422.4	25876.0	61.1850	21,039	19,684	−53.32	188	4201	32.1
C_3H_8	Propane	44.096	−42844.9	−9691.0	61.8660	21,654	19,930	−43.78	183	4081	31.2
C_4H_8	1-Butene	56.107	483.541	29720.4	70.4682	20,839	19,484	−81.40	168	4180	37.1
C_4H_{10}	n-Butane	58.123	−51462.0	−6480.4	71.0628	21,301	19,666	31.10	166	4086	36.1
C_5H_{10}	1-Pentene	70.134	−8631.1	32430.7	79.6749	20,702	19,347	86.00	154	4165	40.0
C_5H_{12}	n-Pentane	72.150	−60417.8	−3383.5	79.8571	21,080	19,499	96.98	154	4090	39.1
C_6H_6	Benzene	78.113	34213.8	53494.0	61.7030	18,176	17,446	176.2	169	4216	54.9
C_6H_{12}	1-Hexene	84.161	−17193.3	35905.4	88.4690	20,617	19,262	146.1	144	4154	42.0
C_6H_{14}	n-Hexane	86.177	−68980.1	86.2287	88.6608	20,936	19,392	156.2	144	4091	41.1
C_7H_{14}	1-Heptene	98.188	−25634.3	39427.1	97.2727	20,558	19,202	200.5	—	4149	—
C_7H_{16}	n-Heptane	100.20	−77490.3	3608.0	97.4742	20,832	19,315	209.1	136	4093	42.7
C_8H_{16}	1-Octene	112.21	−34213.8	42965.8	106.076	20,512	19,157	250.3	—	4144	—
C_8H_{18}	n-Octane	114.23	−86000.5	7146.67	106.278	20,755	19,257	258.3	129	4095	43.9
C_9H_{18}	1-Nonene	126.24	−42706.7	46504.5	114.890	20,478	19,122	—	—	4140	—
C_9H_{20}	n-Nonane	128.26	−94493.5	10668.0	115.082	20,694	19,212	303.4	127	4097	44.8
$C_{10}H_{20}$	1-Decene	140.27	−51217.0	50043.2	123.693	20,449	19,094	339.1	—	4136	—
$C_{10}H_{22}$	n-Decane	142.28	−103004	14206.7	123.895	20,645	19,175	345.4	119	4099	45.6
$C_{11}H_{22}$	1-Undecene	154.30	−59727.2	53564.9	132.497	20,426	19,071	—	—	4133	—
$C_{11}H_{24}$	n-Undecane	156.31	−111514	17745.8	132.699	20,604	19,145	384.6	114	4099	46.2
$C_{12}H_{24}$	1-Dodecene	168.32	−68220.5	57103.6	141.301	20,408	19,052	416.1	—	4131	—
$C_{12}H_{26}$	n-Dodecane	170.34	−120539	—	—	20,568	19,117	421.3	110	4099	46.8

*Based on gaseous fuel.

†For stoichiometric combustion with air (79% N_2, 21% O_2).

‡For liquids at 20°C or for gases at the boiling point of the liquefied gas.

Sources:

[a]Rossini, F. D., et al., *Selected Values of Physical and Thermodynamic Properties of Hydrocarbons and Related Compounds,* Carnegie Press, Pittsburgh, PA, 1953.

[b]Weast, R. C. (Ed.), *Handbook of Chemistry and Physics,* 56th ed., CRC Press, Cleveland, OH, 1976.

[c]Obert, E. F., *Internal Combustion Engines and Air Pollution,* Harper & Row, New York, 1973.

[d]Turns, S. R., *An Introduction to Combustion,* 2nd ed., McGraw-Hill, New York, 2000.

Appendix IE
Thermo-Physical Properties of Selected Solids

Table IE.1 Thermo-Physical Properties of Selected Metallic Solids[a]

Composition	Melting Point (R)	Properties at 540 R				k (Btu/hr·ft·R) and c_p (Btu/lbm·R) at Various Temperatures (R)									
		ρ (lbm/ft³)	c_p (Btu/lbm·R)	k (Btu/hr·ft·R)	$\alpha \cdot 10^6$ (ft²/s)	180	360	720	1080	1440	1800	2160	2700	3600	4500
Aluminum															
Pure	1679	168.7	0.216	136.9	1045	175	137	139	133	126					
						0.115	0.191	0.227	0.247	0.274					
Alloy 2024-T6 (4.5% Cu, 1.5% Mg, 0.6% Mn)	1395	172.9	0.209	102.3	786	37.6	94.2	107	107						
Alloy 195, Cast (4.5% Cu)		174.2	0.211	97.1	734			101	107						
						0.113	0.188	0.221	0.249						
Beryllium	2790	115.5	0.436	115.6	637	572	174	93.0	72.8	61.2	52.5	45.5			
						0.0485	0.266	0.523	0.622	0.674	0.721	0.771	0.841		
Bismuth	981.0	610.5	0.029	4.5	71	9.53	5.60	4.07							
						0.0268	0.0287	0.0303							
Boron	4631	156.1	0.264	15.6	105	110	32.1	9.71	6.12	5.55	5.69				
						0.0306	0.143	0.349	0.452	0.516	0.558				
Cadmium	1069	540.0	0.055	55.9	521	117	57.4	54.7							
						0.0473	0.0530	0.0578							
Chromium	3812	447.0	0.107	54.1	313	91.9	64.1	52.5	46.6	41.2	37.8	35.8	33.1	28.5	
						0.0459	0.0917	0.116	0.129	0.139	0.147	0.163	0.186	0.224	
Cobalt	3184	553.2	0.101	57.3	286	96.5	70.5	49.3	38.9	33.6	30.1	28.5	24.6		
						0.0564	0.0905	0.107	0.120	0.131	0.150	0.175	0.161		
Copper															
Pure	2444	557.7	0.092	231.7	1259	279	239	227	219	211	203	196			
						0.0602	0.0850	0.0948	0.0996	0.103	0.108	0.115			
Commercial bronze (90% Cu, 10% Al)	2327	549.4	0.100	30.0	151		24.3	30.0	34.1						
Phosphor gear bronze (89% Cu, 11% Sn)	1987	548.1	0.085	31.2	183		23.7	37.6	42.8						
							0.187	0.110	0.130						
Cartridge brass (70% Cu, 30% Zn)	2138	532.5	0.091	63.6	365	43.3	54.9	79.2	86.1						
						0.0566	0.0860	0.0943	0.102						
Constantan (55% Cu, 45% Ni)	2687	556.9	0.092	13.3	72	9.82	11.0								
							0.0865								
Germanium	2180	334.6	0.077	34.6	374	134	55.9	25.0	15.8	11.4	10.1	10.1			
						0.0454	0.0693	0.0805	0.0831	0.0853	0.0896	0.0943			
Gold	2405	1204.8	0.031	183.2	1367	189	187	180	172	164	156	147			
						0.0260	0.0296	0.0313	0.0322	0.0334	0.0346	0.0370			
Iridium	4896	1404.6	0.031	84.9	541	99.4	88.4	83.2	79.7	76.3	72.8	69.3	64.1		
						0.0215	0.0291	0.0318	0.0330	0.0344	0.0365	0.0385	0.0411		
Iron															
Pure	3258	491.3	0.107	46.3	249	77.4	54.3	40.2	31.6	25.0	19.0	16.4	18.5		
						0.0516	0.0917	0.117	0.137	0.162	0.233	0.145	0.156		
Armco (99.75% pure)		491.3	0.107	42.0	223	55.2	46.6	38.0	30.7	24.4	18.7	16.6	18.1		
						0.0514	0.0917	0.117	0.137	0.162	0.233	0.145	0.156		

(continued)

Composition	Melting Point	ρ	c_p	k	α	k (c_p) at successive temperatures →
Iron						
Wrought iron* (C <0.5%)		490.0	0.110	34.1	175	48.0, 34.7, 32.4, 27.2, 22.5, 19.6, 19.1, 19.1
Cast iron* (C ≈ 4%)		454.0	0.100	30.0	183	
Carbon steels						
Carbon steel* (C ≈ 0.5%)		489.0	0.111	31.2	158	32.9, 29.5, 25.4, 22.0, 18.5, 17.3, 17.9
Carbon steel* (C ≈ 1.0%)		487.0	0.113	24.8	126	24.8, 24.8, 22.5, 19.6, 17.3, 16.2, 16.8
Carbon steel* (C ≈ 1.5%)		484.0	0.116	20.8	104	20.8, 20.8, 19.6, 18.5, 16.8, 16.2, 16.8
Carbon steels						
Plain carbon (Mn ≤ 1%, Si ≤ 0.1%)		490.3	0.104	35.0	191	32.8 (0.116), 27.7 (0.134), 22.7 (0.164), 17.3 (0.279)
AISI 1010		488.9	0.104	36.9	202	33.9 (0.116), 28.2 (0.134), 22.7 (0.164), 18.1 (0.279)
Carbon–silicon (Mn ≤ 1%, 0.1% < Si ≤ 0.6%)		488.0	0.107	30.0	160	28.8 (0.116), 25.4 (0.134), 21.6 (0.164), 16.9 (0.279)
Carbon–manganese–silicon (1% < Mn ≤ 1.65%, 0.1% < Si ≤ 0.6%)		507.6	0.104	23.7	125	24.4 (0.120), 22.9 (0.139), 20.2 (0.167), 15.9 (0.232)
Chromium (low) steels						
1/2 Cr–1/4 Mo–Si (0.18% C, 0.65% Cr, 0.23% Mo, 0.6% Si)		488.3	0.106	21.8	117	22.1 (0.116), 21.2 (0.134), 19.2 (0.164), 15.5 (0.260)
1 Cr–1/2 Mo (0.16% C, 1% Cr, 0.54% Mo, 0.39% Si)		490.6	0.106	24.4	131	24.3 (0.118), 22.6 (0.008), 19.9 (0.164), 15.8 (0.231)
1 Cr–V (0.2% C, 1.02% Cr, 0.15% V)		489.2	0.106	28.3	152	27.0 (0.118), 24.3 (0.137), 21.0 (0.164), 16.3 (0.231)
Stainless steels AISI 302		502.9	0.115	8.7	42	10.00 (0.122), 11.6 (0.134), 13.2 (0.140), 14.7 (0.145); (0.118)
AISI 304	3006	493.2	0.114	8.6	43	5.32 (0.0650), 7.28 (0.0960), 9.59 (0.123), 11.4 (0.133), 13.1 (0.139), 14.7 (0.146), 16.2 (0.153), 18.3 (0.163)
AISI 316		514.3	0.112	7.7	37	8.78 (0.120), 10.6 (0.131), 12.3 (0.138), 14.0 (0.144)
AISI 347		498.0	0.115	8.2	40	9.13 (0.123), 10.9 (0.134), 12.7 (0.140), 14.3 (0.145)
Lead	1082	707.9	0.031	20.4	259	22.9 (0.0282), 21.2 (0.0299), 19.6 (0.0315), 18.1 (0.0339)
Magnesium	1661	108.6	0.245	90.1	943	97.7 (0.155), 91.9 (0.223), 88.4 (0.257), 86.1 (0.279), 84.4 (0.303)
Molybdenum	5209	639.3	0.060	79.7	578	103 (0.0337), 82.6 (0.0535), 77.4 (0.0623), 72.8 (0.0657), 68.2 (0.0681), 64.7 (0.0705), 60.7 (0.0736), 56.6 (0.0788), 52.0 (0.0908), 49.7 (0.110)
Nickel						
Pure	3110	555.6	0.106	52.4	248	94.8 (0.0554), 61.8 (0.0915), 46.3 (0.116), 41.5 (0.134), 39.1 (0.127), 37.9 (0.141), 44.0 (0.142), 47.7 (0.147)
Nichrome (80% Ni, 20% Cr)	3010	524.4	0.100	6.9	37	8.09 (0.115), 9.25 (0.125), 12.1 (0.130)

Table IE.1 (continued)

In the "Various Temperatures" columns each cell lists k (Btu/hr·ft·R) as the upper value and c_p (Btu/lbm·R) as the lower value.

Composition	Melting Point (R)	Properties at 540 R — ρ (lbm/ft³)	c_p (Btu/lbm·R)	k (Btu/hr·ft·R)	$\alpha \cdot 10^6$ (ft²/s)	180	360	720	1080	1440	1800	2160	2700	3600	4500
Inconel X-750 (73% Ni, 15% Cr, 6.7% Fe)	2997	531.3	0.105	6.8	33	5.03	5.95	7.80	9.82	11.8	13.9	15.9	19.1		
Niobium	4934	535.0	0.063	31.0	254	— / 0.0449	30.4 / 0.0595	31.9 / 0.0654	33.6 / 0.0676	35.4 / 0.0697	37.2 / 0.0719	39.0 / 0.0740	41.7 / 0.0774	45.7 / 0.0829	
Palladium	3289	750.4	0.058	41.5	264	44.2 / 0.0401	41.4 / 0.0542	42.5 / 0.0600	46.1 / 0.0623	50.2 / 0.0647	54.4 / 0.0671	58.9 / 0.0695	63.6 / 0.0733		
Platinum Pure	3681	1339.1	0.032	41.4	270	44.8 / 0.0239	41.9 / 0.0299	41.5 / 0.0325	42.3 / 0.0337	43.7 / 0.0349	45.5 / 0.0363	47.7 / 0.0375	51.7 / 0.0394	57.4 / 0.0428	
Alloy 60Pt–40Rh (60% Pt, 40% Rh)	3240	1038.2	0.039	27.2	187			30.0	34.1	37.6	39.9	42.2	43.9		
Rhenium	6215	1317.2	0.032	27.7	180	34.0 / 0.0232	29.5 / 0.0303	26.6 / 0.0332	25.5 / 0.0346	25.5 / 0.0361	25.8 / 0.0373	26.4 / 0.0387	27.6 / 0.0408	30.0 / 0.0444	
Rhodium	4025	777.2	0.058	86.7	534	107 / 0.0351	89.0 / 0.0525	84.4 / 0.0604	78.6 / 0.0654	73.4 / 0.0700	69.9 / 0.0743	67.0 / 0.0781	63.6 / 0.0834	64.7 / 0.0898	
Silicon	3033	145.5	0.170	85.5	960	511 / 0.0619	153 / 0.1328	57.1 / 0.189	35.8 / 0.207	24.4 / 0.218	18.0 / 0.226	14.8 / 0.231	13.1 / 0.237		
Silver	2223	655.5	0.056	247.9	1873	257 / 0.0447	248 / 0.0537	246 / 0.0571	238 / 0.0597	229 / 0.0626	219 / 0.0662	209 / 0.0697			
Tantalum	5884	1036.3	0.033	33.2	266	34.2 / 0.0263	33.2 / 0.0318	33.4 / 0.0344	33.9 / 0.0349	34.3 / 0.0356	34.8 / 0.0363	35.2 / 0.0370	35.9 / 0.0382	37.0 / 0.0411	37.9 / 0.0451
Thorium	3641	730.4	0.028	31.2	421	34.6 / 0.0236	31.5 / 0.0268	31.5 / 0.0296	32.2 / 0.0320	32.9 / 0.0346	32.9 / 0.0373	33.9 / 0.0399			
Tin	909	456.3	0.054	38.5	432	49.2 / 0.0449	42.4 / 0.0514	35.9 / 0.0580							
Titanium	3515	280.9	0.125	12.7	100	17.6 / 0.0717	14.2 / 0.111	11.8 / 0.132	11.2 / 0.141	11.4 / 0.151	12.0 / 0.161	12.7 / 0.148	14.2 / 0.164		
Tungsten	6588	1204.8	0.032	100.5	735	120 / 0.0208	107 / 0.0291	91.9 / 0.0327	79.2 / 0.0339	72.2 / 0.0346	68.2 / 0.0353	65.3 / 0.0363	61.8 / 0.0375	57.8 / 0.0399	54.9 / 0.0420
Uranium	2531	1190.5	0.028	15.9	135	12.5 / 0.0225	14.5 / 0.0258	17.1 / 0.0299	19.6 / 0.0349	22.4 / 0.0420	25.4 / 0.0430	28.3 / 0.0385			
Vanadium	3946	380.8	0.117	17.7	111	20.7 / 0.0616	18.1 / 0.103	18.1 / 0.123	19.2 / 0.129	20.6 / 0.134	22.1 / 0.143	23.6 / 0.154	25.8 / 0.171	29.4 / 0.207	
Zinc	1247	445.7	0.093	67.0	450	67.6 / 0.0709	68.2 / 0.0877	64.1 / 0.0960	59.5 / 0.104						
Zirconium	3825	410.1	0.066	13.1	133	19.2 / 0.0490	14.6 / 0.0631	12.5 / 0.0717	12.0 / 0.0769	12.5 / 0.0817	13.7 / 0.0865	15.0 / 0.0822	16.6 / 0.0822	19.1 / 0.0822	

Source: [a]Adapted from Incropera, F. P., and DeWitt, D. P., *Fundamentals of Heat and Mass Transfer*, 3rd ed., Wiley, New York, 1990, with permission. Data for wrought iron, cast iron, and various carbon steels adapted from Chapman, A. J., *Fundamentals of Heat Transfer*, Macmillan, New York, 1987.

Table IE.2 Thermo-Physical Properties of Selected Nonmetallic Solids[a]

Column groups: **Properties at 540 R** spans ρ, c_p, k, α. **k (Btu/hr·ft·R) and c_p (Btu/lbm·R) at Various Temperatures (R)** spans the columns 180 through 3600. In the temperature columns, the first (upper) value of each material is k and the row labeled c_p gives the specific heat.

Composition	Melting Point (R)	ρ (lbm/ft³)	c_p (Btu/lbm·R)	k (Btu/hr·ft·R)	$\alpha \cdot 10^6$ (ft²/s)	180	360	720	1080	1440	1800	2160	2700	3600
Aluminum oxide, sapphire	4181	247.8	0.183	26.6	162.5	260	47.4	18.7	10.9	7.5	6.1			
c_p								0.225	0.265	0.282	0.293			
Aluminum oxide, polycrystalline	4181	247.8	0.183	20.8	128.1	76.9	31.8	15.3	9.13	6.01	4.54	3.78	3.27	3.47
c_p								0.225	0.265	0.282	0.293			
Beryllium oxide	4905	187.3	0.246	157.2	947.2	—	—	113	64.1	40.4	27.2	19.1	12.4	8.67
c_p								0.322	0.404	0.445	0.472	0.491	0.512	0.657
Boron	4631	156.1	0.264	15.9	107.5	110	30.3	10.8	6.53	4.68	3.64	3.00		
c_p								0.356	0.449	0.510	0.561	0.610		
Boron fiber epoxy (30% vol) composite	1062	129.8												
k, ‖ to fibers				1.32		1.21	1.29	1.32						
k, ⊥ to fibers				0.34		0.21	0.28	0.35						
c_p			0.268			0.087	0.181	0.342						
Carbon, Amorphous	2700	121.7	—	0.92	—	0.39	0.68	1.09	1.27	1.37	1.46	1.64	2.01	
Diamond, type IIa insulator	—	218.5	0.122	1329	—	5778	2311	890						
c_p						0.005	0.046	0.204						
Graphite, pyrolytic	4091	138.0												
k, ‖ to layers				1127		2872	1866	803	515	385	309	259	206	151
k, ⊥ to layers				3.29		9.71	5.33	2.36	1.55	1.16	0.92	0.77	0.62	0.47
c_p			0.169			0.032	0.098	0.237	0.336	0.394	0.428	0.451	0.471	0.488
Graphite fiber epoxy (25% vol) composite	810.0	87.4												
k, heat flow ‖ to fibers				6.41		3.29	5.03	7.51						
k, heat flow ⊥ to fibers				0.50		0.27	0.39	0.64						
c_p			0.223			0.080	0.153	0.290						
Pyroceram, Corning 9606	2921	162.3	0.193	2.30	20.3	3.03	2.76	2.10	1.90	1.78	1.71	1.66	1.61	
c_p								0.217	0.248	0.268	0.286	0.302	0.358	
Silicon carbide	5580	197.3	0.161	283.1	2476	—	—				50.3	33.5	17.3	
c_p								0.210	0.251	0.271	0.285	0.297	0.313	
Silicon dioxide, crystalline (quartz)	3389	165.4												
k, ‖ to c axis				6.01		22.5	9.48	4.39	2.89	2.43				
k, ⊥ to c axis				3.59		12.0	5.49	2.72	1.96	1.79				
c_p			0.178					0.211	0.257	0.299				
Silicon dioxide, polycrystalline (fused silica)	3389	138.6	0.178	0.80	8.98	0.40	0.66	0.87	1.01	1.25	1.66	2.31		
c_p							0.138	0.216	0.248	0.264	0.276	0.285		
Silicon nitride	3911	149.8	0.165	9.25	103.9	—	—	8.03	6.53	5.71	5.06	4.62	4.14	3.58
c_p								0.186	0.224	0.254	0.276	0.293	0.312	0.329
Sulfur	705.6	129.2	0.169	0.12	1.52	0.10	0.11							
c_p						0.096	0.145							
Thorium dioxide	6431	568.7	0.056	7.51	65.7			5.89	3.81	2.72	2.13	1.80	1.58	1.44
c_p								0.061	0.065	0.068	0.070	0.072	0.075	0.079
Titanium dioxide, polycrystalline	3839	259.5	0.170	4.85	30.1			4.05	2.90	2.28	2.00	1.90		
c_p								0.192	0.210	0.217	0.222	0.226		

Source: [a]Adapted from Incropera, F. P., and Dewitt, D. P., *Fundamentals of Heat and Mass Transfer*, 3rd ed., Wiley, New York, 1990, with permission.

Table IE.3   Thermo-Physical Properties of Common Materials[a]

Description/ Composition	Temperature (R)	Density, ρ (lbm/ft^3)	Thermal Conductivity, k (Btu/hr·ft·R)	Specific Heat, c_p (Btu/lbm·R)
Asphalt	540	132.0	0.036	0.220
Bakelite	540	81.16	0.809	0.350
Brick, refractory				
Carborundum	1570	—	10.69	—
	3010	—	6.356	—
Chrome brick	851	187.9	1.329	0.199
	1481	—	1.445	—
	2111	—	1.156	—
Diatomaceous	860	—	0.144	—
silica, fired	2061	—	0.173	—
Fire clay, burnt 1600 K	1391	128.0	0.578	0.229
	1931	—	0.636	—
	2471	—	0.636	—
Fire clay, burnt 1725 K	1391	145.1	0.751	0.229
	1931	—	0.809	—
	2471	—	0.809	—
Fire clay brick	860	165.1	0.578	0.229
	1660	—	0.867	—
	2660	—	1.040	—
Magnesite	860	—	2.196	0.270
	1660	—	1.618	—
	2660	—	1.098	—
Clay	540	91.14	0.751	0.210
Coal, anthracite	540	84.28	0.150	0.301
Concrete (stone mix)	540	143.6	0.809	0.210
Cotton	540	4.994	0.035	0.311
Foodstuffs				
Banana (75.7% water content)	540	61.18	0.278	0.800
Apple, red (75% water content)	540	52.44	0.296	0.860
Cake, batter	540	44.95	0.129	—
Cake, fully baked	540	17.48	0.070	—
Chicken meat, white	356	—	0.925	—
(74.4% water content)	419	—	0.861	—
	455	—	0.780	—
	473	—	0.693	—
	491	—	0.275	—
	509	—	0.277	—
	527	—	0.283	—
Glass				
Plate (soda lime)	540	156.1	0.809	0.179
Pyrex	540	138.9	0.809	0.199
Ice	491	57.43	1.086	0.487
	455	—	1.173	0.465
Leather (sole)	540	62.30	0.092	—
Paper	540	58.06	0.104	0.320
Paraffin	540	56.18	0.139	0.690
Rock				
Granite, Barre	540	164.2	1.612	0.185
Limestone, Salem	540	144.8	1.242	0.193
Marble, Halston	540	167.3	1.618	0.198

Description/ Composition	Temperature (R)	Density, ρ (lbm/ft^3)	Thermal Conductivity, k(Btu/hr·ft·R)	Specific Heat, c_p (Btu/lbm·R)
Quartzite, Sioux	540	164.8	3.109	0.264
Sandstone, Berea	540	134.2	1.676	0.178
Rubber, vulcanized				
Soft	540	68.67	0.075	0.480
Hard	540	74.29	0.092	—
Sand	540	94.58	0.156	0.191
Soil	540	128.0	0.300	0.439
Snow	491	6.867	0.028	—
		31.21	0.110	—
Teflon	540	137.3	0.202	—
	720	—	0.260	—
Tissue, human				
Skin	540	—	0.214	—
Fat layer (adipose)	540	—	0.116	—
Muscle	540	—	0.237	—
Wood, cross grain				
Balsa	540	8.740	0.032	—
Cypress	540	29.03	0.056	—
Fir	540	25.91	0.064	0.650
Oak	540	34.02	0.098	0.570
Yellow pine	540	39.95	0.087	0.670
White pine	540	27.16	0.064	—
Wood, radial				
Oak	540	34.02	0.110	0.570
Fir	540	26.22	0.081	0.650

Source: [a]Adapted from Incropera, F. P., and DeWitt, D. P., *Fundamentals of Heat and Mass Transfer,* 3rd ed., Wiley, New York, 1990, with permission.

Table IE.4   Thermo-Physical Properties of Structural Building Materials[a]

Description/Composition	Density, ρ (lbm/ft^3)	Typical Properties at 540 R Thermal Conductivity, k (Btu/hr·ft·R)	Specific Heat, c_p (Btu/lbm·R)
Building boards			
Asbestos–cement board	120	0.335	—
Gypsum or plaster board	50	0.098	—
Plywood	34	0.069	0.290
Sheathing, regular density	18	0.032	0.311
Acoustic tile	18	0.034	0.320
Hardboard, siding	40	0.054	0.279
Hardboard, high density	63	0.087	0.330
Particle board, low density	37	0.045	0.311
Particle board, high density	62	0.098	0.311
Woods			
Hardwoods (oak, maple)	45	0.092	0.300
Softwoods (fir, pine)	32	0.069	0.330
Masonry materials			
Cement mortar	116	0.416	0.186
Brick, common	120	0.416	0.199
Brick, face	130	0.751	—
Clay tile, hollow			
1 cell deep, 10 cm thick	—	0.300	—
3 cells deep, 30 cm thick	—	0.399	—
Concrete block, 3 oval cores			
sand/gravel, 20 cm thick	—	0.578	—
cinder aggregate, 20 cm thick	—	0.387	—
Concrete block, rectangular core			
2 cores, 20 cm thick, 16 kg	—	0.636	—
same with filled cores	—	0.347	—
Plastering materials			
Cement plaster, sand aggregate	116	0.416	—
Gypsum plaster, sand aggregate	105	0.127	0.259
Gypsum plaster, vermiculite aggregate	45	0.144	—
Blanket and batt			
Glass fiber, paper faced	1.0	0.027	—
	1.7	0.022	—
	2.5	0.020	—
Glass fiber, coated; duct liner	2.0	0.022	0.199
Board and slab			
Cellular glass	9.1	0.034	0.239
Glass fiber, organic bonded	6.6	0.021	0.190
Polystyrene, expanded			
extruded (R-12)	3.4	0.016	0.289
molded beads	1.0	0.023	0.289
Mineral fiberboard; roofing material	17	0.028	—
Wood, shredded/cemented	22	0.050	0.380
Cork	7.5	0.023	0.430
Loose fill			
Cork, granulated	10	0.026	—
Diatomaceous silica, coarse powder	22	0.040	—
	25	0.053	—
Diatomaceous silica, fine powder	12	0.030	—
	17	0.035	—

Description/Composition	Typical Properties at 540 R		
	Density, ρ (lbm/ft^3)	Thermal Conductivity, k (Btu/hr·ft·R)	Specific Heat, c_p (Btu/lbm·R)
Glass, fiber, poured or blown	1.0	0.025	0.199
Vermiculite, flakes	5.0	0.039	0.199
	10	0.036	0.239
Formed/foamed-in-place			
Mineral wool granules with asbestos/inorganic binders, sprayed	12	0.027	—
Polyvinyl acetate cork mastic; sprayed or troweled	—	0.058	—
Urethane, two-part mixture; rigid foam	4.4	0.015	0.250
Reflective			
Aluminum foil separating fluffy glass mats; 10–12 layers; evacuated; for cryogenic applications (150 K)	2.5	0.0000925	—
Aluminum foil and glass paper laminate; 75–150 layers; evacuated; for cryogenic application (150 K)	7.5	0.0000098	—
Typical silica powder, evacuated	10	0.001	—

Source: [a]Adapted from Incropera, F. P., and DeWitt, D. P., *Fundamentals of Heat and Mass Transfer,* 3rd ed., Wiley, New York, 1990, with permission.

Table IE.5 Thermo-Physical Properties of Industrial Insulation[a]

Description/Composition	Max Service Temp (R)	Typical Density (lbm/ft³)	Typical Thermal Conductivity k (Btu/hr·ft·R) at Various Temperatures (R)													
			360	387	414	432	459	486	513	540	558	657	756	954	1161	1350
Blankets																
Blanket, mineral fiber, metal reinforced	1656	6.00–12.0									0.022	0.027	0.032	0.045		
	1467	2.50–6.00									0.020	0.026	0.034	0.051		
Blanket, mineral fiber, glass; fine fiber, organic bonded	810	0.624				0.021	0.022	0.023	0.025	0.028	0.030	0.044				
		0.749				0.020	0.021	0.023	0.024	0.027	0.028	0.040				
		0.999				0.019	0.020	0.021	0.023	0.024	0.027	0.036				
		1.50				0.017	0.018	0.019	0.021	0.023	0.023	0.031				
		2.00				0.017	0.017	0.018	0.019	0.021	0.022	0.028				
		3.00				0.016	0.017	0.017	0.018	0.019	0.020	0.026				
Blanket, alumina–silica fiber	2754	3.00												0.041	0.061	0.087
		4.00												0.034	0.050	0.072
		5.99												0.030	0.044	0.058
		7.99												0.028	0.039	0.053
Felt, semirigid; organic bonded	864	3.12–7.80	0.013	0.014	0.015	0.016	0.017	0.020	0.021	0.022	0.023	0.029	0.036			
Felt, laminated; no binder	1314	3.12						0.017	0.018	0.019	0.020	0.029	0.046			
	1656	7.49											0.029	0.038	0.050	
Blocks, boards, and pipe insulations																
Asbestos paper, laminated and corrugated																
4-ply	756	11.9								0.045	0.047	0.057				
6-ply	756	15.9								0.041	0.043	0.049				
8-ply	756	18.7								0.039	0.041	0.047				

Material														
Magnesia, 85%	1062	11.5						0.029	0.032	0.035	0.043	0.051	0.060
Calcium silicate	1656	11.9						0.032	0.034	0.036			
Cellular glass	1260	9.05	0.027	0.028	0.029	0.030	0.032	0.034	0.036	0.040	0.046		
Diatomaceous	2061	21.5								0.053	0.057	0.060	
silica	2358	24.0								0.058	0.058	0.066	
Polystyrene, rigid													
Extruded (R-12)	630	3.50	0.013	0.013	0.013	0.013	0.014	0.015	0.016	0.017			
Extruded (R-12)	630	2.18		0.013	0.013	0.014	0.014	0.015	0.016	0.017			
Molded beads	630	0.999	0.015	0.017	0.017	0.019	0.020	0.021	0.022	0.023			
Rubber, rigid foamed	612	4.37					0.029	0.029	0.029	0.029			
Insulating cement													
Mineral fiber (rock, slag, or glass)													
with clay binder	2259	26.8							0.041	0.046	0.051	0.061	0.071
with hydraulic setting binder	1660	35.0							0.062	0.066	0.071	0.079	
Loose fill													
Cellulose, wood, or paper pulp	—	2.81	0.021	0.023	0.024	0.025	0.027	0.028	0.022	0.023	0.024		
Perlite, expanded	—	6.55	0.032	0.034	0.035	0.034	0.035	0.029	0.031	0.032			
Vermiculite,	—	7.62				0.032	0.036	0.038	0.039	0.041			
expanded		4.99	0.028	0.029	0.032	0.034	0.035	0.036	0.038				

Source: [a]Adapted from Incropera, F. P., and DeWitt, D. P., *Fundamentals of Heat and Mass Transfer*, 3rd ed., Wiley, New York, 1990, with permission.

Appendix LE
Psychrometry Chart

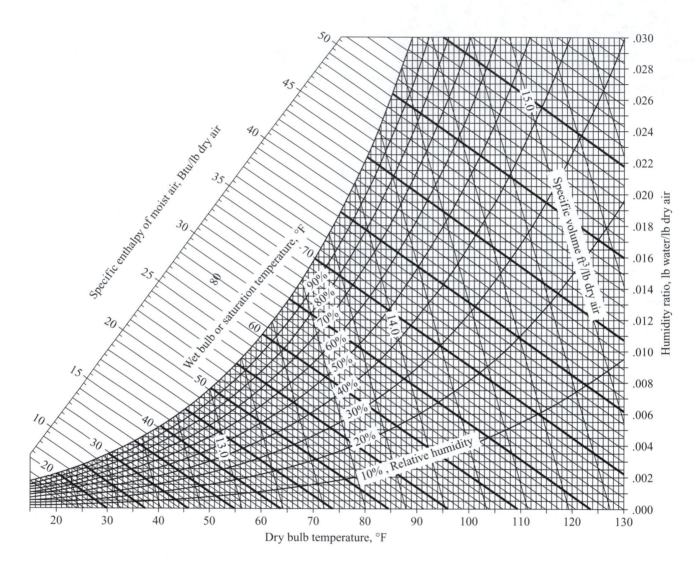

FIGURE LE.1
Psychrometric chart in U. S. customary units (P = 14.7 psia). Adapted with permission from Z. Zhang and M. B. Pate, "A Methodology for Implementing a Psychrometric Chart in a Computer Graphics System," *ASHRAE Transactions*, Vol. 94, Pt. 1, 1988.

Appendix ME
Properties of the Atmosphere at High Altitude

Table ME.1   Properties of the Atmosphere at High Altitude*

Altitude (m)	Temperature (°C)	Pressure (kPa)	Density (kg/m³)	Acceleration due to Gravity (m/s²)	Speed of Sound (m/s)	Viscosity (N·s/m²)	Thermal Conductivity (W/m·K)
0	59.0	14.695	0.076474	32.174	1116.4	0.000012024	0.014634
500	57.2	14.433	0.075363	32.173	1114.5	0.000011992	0.014589
1000	55.4	14.173	0.074264	32.171	1112.6	0.000011960	0.014544
1500	53.7	13.917	0.073172	32.169	1110.6	0.000011927	0.014499
2000	51.9	13.665	0.072098	32.168	1108.7	0.000011895	0.014454
2500	50.1	13.416	0.071037	32.166	1106.8	0.000011863	0.014409
3000	48.3	13.172	0.069982	32.165	1104.9	0.000011831	0.014363
3500	46.5	12.930	0.068945	32.163	1102.9	0.000011798	0.014318
4000	44.7	12.693	0.067915	32.162	1101.0	0.000011766	0.014273
4500	43.0	12.459	0.066904	32.160	1099.0	0.000011733	0.014227
5000	41.2	12.228	0.065899	32.158	1097.1	0.000011700	0.014182
5500	39.4	12.001	0.064906	32.157	1095.1	0.000011667	0.014136
6000	37.6	11.778	0.063926	32.156	1093.2	0.000011635	0.014091
6500	35.8	11.558	0.062959	32.154	1091.2	0.000011602	0.014045
7000	34.1	11.341	0.061998	32.153	1089.3	0.000011569	0.014000
7500	32.3	11.127	0.061051	32.151	1087.3	0.000011536	0.013954
8000	33.5	10.917	0.060116	32.149	1085.3	0.000011503	0.013909
8500	28.7	10.710	0.059191	32.148	1083.3	0.000011471	0.013863
9000	26.9	10.507	0.058278	32.146	1081.4	0.000011438	0.013817
9500	25.1	10.306	0.057375	32.145	1079.4	0.000011404	0.013771
10000	23.4	10.109	0.056483	32.143	1077.4	0.000011371	0.013726
11000	19.8	9.7228	0.054731	32.140	1073.4	0.000011305	0.013634
12000	16.2	9.3490	0.053021	32.137	1069.4	0.000011238	0.013542
13000	12.7	8.9871	0.051353	32.134	1065.4	0.000011171	0.013450
14000	9.10	8.6366	0.049725	32.131	1061.4	0.000011104	0.013357
15000	5.54	8.2973	0.048138	32.128	1057.3	0.000011036	0.013265
16000	1.99	7.9689	0.046589	32.125	1053.3	0.000010969	0.013173
17000	21.57	7.6512	0.045079	32.122	1049.2	0.000010901	0.013080
18000	25.13	7.3437	0.043607	32.118	1045.1	0.000010833	0.012987
19000	28.70	7.0465	0.042171	32.115	1041.0	0.000010764	0.012893
20000	212.3	6.7590	0.040773	32.113	1036.9	0.000010696	0.012800
22000	219.4	6.2126	0.038083	32.106	1028.6	0.000010558	0.012613
24000	226.5	5.7027	0.035532	32.100	1020.3	0.000010419	0.012425
26000	233.6	5.2272	0.033113	32.094	1011.9	0.000010279	0.012237
28000	240.7	4.7846	0.030823	32.088	1003.4	0.000010138	0.012048
30000	247.8	4.3728	0.028657	32.082	994.8	0.000009996	0.011858
32000	254.9	3.9902	0.026609	32.075	986.2	0.000009853	0.011668
34000	262.1	3.6353	0.024677	32.070	977.5	0.000009709	0.011477
36000	269.2	3.3065	0.022853	32.063	968.7	0.000009564	0.011285
38000	269.7	3.0045	0.020794	32.057	968.1	0.000009553	0.011270
40000	269.7	2.7301	0.018895	32.051	968.1	0.000009553	0.011270
45000	269.7	2.1491	0.014873	32.036	968.1	0.000009553	0.011270
50000	269.7	1.6917	0.011709	32.020	968.1	0.000009553	0.011270
55000	269.7	1.3320	0.0092187	32.005	968.1	0.000009553	0.011270
60000	269.7	1.0488	0.0072591	31.990	968.1	0.000009553	0.011270
65000	269.7	0.82594	0.0057164	31.974	968.1	0.000009553	0.011270

*United States Committee on Extension to the Standard Atmosphere, "U.S. Standard Atmosphere, 1976", National Oceanic and Atmospheric Administration, National Aeronautics and Space Administration, United States Air Force, Washington D.C., 1976.